Special Operations and National Purpose

Special Operations and National Purpose

By

Ross S. Kelly

Lexington Books

D.C. Heath and Company/Lexington, Massachusetts/Toronto

Library of Congress Cataloging-in-Publication Data

Kelly, Ross S.
Special operations and national purpose.

Includes index
1. Special operations (Military science). 2. Special forces (Military science)—United States. 3. Special forces (Military science)—Europe. 4. Special forces (Military science)—Soviet Union. I. Title.
U262.K45 1989 355'.0218 87-46111
ISBN 0-669-17410-6 (alk. paper)

Published simultaneously in Canada
Printed in the United States of America
International Standard Book Number: 0-669-17410-6
Library of Congress Catalog Card Number: 87-46111

The paper used in this publication meets the minimum requirements of American National Standard for Information Sciences—Permanence of Paper for Printed Library Materials, ANSI Z39.48-1984.

ISBN 0-669-17410-6

89 90 91 92 8 7 6 5 4 3 2 1

To all
"soldiers of the shadows" past and present,
and to my wife, Susan,
who married one

Contents

Foreword

More than a quarter century ago, President John F. Kennedy warned, "The Free World's security can be endangered not only by a nuclear attack, but also by being slowly nibbled away at the periphery, regardless of our strategic power, by forces of subversion, infiltration, intimidation, indirect or non-overt aggression, internal revolution, diplomatic blackmail, guerrilla warfare, or a series of limited wars." Despite the seriousness of this threat, there has been an overwhelming emphasis on conventional and strategic military forces by U.S. defense planners and their political allies, a focus on the war that may never come rather than the war that is already here. By contrast, low-intensity conflict (LIC)—terrorism, coups d'état and revolution, guerrilla warfare, paramilitary covert operations, raids and sabotage, psychological operations—has received little systematic attention until recently. This is partially a legacy of the Vietnam War and the belief, in some quarters, that the development of expanded LIC capabilities will lead to a new U.S. interventionism in the affairs of other nations.

The sorry state of U.S. LIC capabilities that the Reagan administration found upon taking office in 1981 was also the product of a disastrous alliance between senior Pentagon planners and the defense industry, who were chiefly interested in promoting military expenditures for complex weapons systems and other high-ticket items. LIC had few advocates, in part, because there was no money in it. By contrast to conventional forces, which are highly dependent on hardware and technology, unconventional warfare places a premium on mobility, leanness, and superior training and motivation. It emphasizes warriors instead of military managers, small unit tactics instead of set-piece battles, innovation over doctrine, and operations conducted in peacetime as well as in time of war.

Fortunately, there has been a new recognition of the need to expand and revitalize America's special operations forces to meet the growing threat of international terrorism and the Soviet effort to destabilize the developing world, often through proxies and surrogates. The notion that unconventional responses are needed to meet unconventional threats has gained greater currency in Washington, even among some lawmakers not known for their advocacy of the defense community. Recent years have witnessed new budget outlays for U.S. special operations forces (SOF), the creation of a consolidated special operations command, the development of so-called light brigades, and an increase in the authorized manpower levels for SOF units. Perhaps just as important, SOF officers are finally being promoted to general and flag rank, overcoming the traditional bias that has excluded them for so long from leadership positions in the armed services.

Ross S. Kelly believes that this process of reversing the erosion of the U.S. special warfare capability is already well under way. Kelly describes the effort by the Congress and the Reagan administration to expand and revitalize U.S. special operations capabilities, often in the face of opposition by the Joint Chiefs of Staff and Pentagon traditionalists, to meet the challenges ahead. He examines the special operations capabilities of U.S. allies in Europe, as well as the threat posed by Soviet special purpose forces (*spetsnaz*), which far outnumber their Western counterparts. Kelly pays particular attention to the evolution of Western counterterrorist units, which were virtually nonexistent until the mid-1970s, and to their recent deployment against terrorists. As an indication of how successful Western governments have become in combating terrorism, imprisoned terrorist Alberto Franceschini, one of the founders of the Italian Red Brigades, recently said that terrorist violence no longer pays because Western governments are "stronger and better prepared" than their terrorist adversaries.

This is an important book because of the range of material it covers and the comprehensive nature of the treatment. Kelly has written perhaps the most complete book available today on the subject of special operations. Unlike some other writers in the field, moreover, Kelly has firsthand knowledge about the subject matter.

He is a senior Defense Intelligence Agency official, actively involved in LIC matters, and also serves as a U.S. Special Forces reservist. He has traveled the world visiting special operations units and meeting with allied policymakers regarding mutual security concerns.

Kelly's book is the fourth title to be published as part of the Lexington Books Issues in Low-Intensity Conflict series; more than a half-dozen other books are scheduled for release by 1990. This series represents the continuing commitment of Lexington Books to publishing books that examine the cutting-edge issues of U.S. national security policy. While an effort will be made to publish works of interest to scholars as well as general readers, a special emphasis of the series will be on books of practical value to policymakers and hands-on specialists in the field.

—Neil C. Livingstone

Preface and Acknowledgments

In early 1984, Greg Copley, publisher of *Defense & Foreign Affairs,* and I met to discuss the quantitative and qualitative upgrades to special operations capabilities taking place in the United States, the Soviet Union, and Europe. We agreed that it was a significant phenomenon in international security affairs, deserving of informed coverage and assessment. Out of that discussion came "Special Operations Report," a series of columns on special operations and low intensity conflict that concluded in 1987. Copley recognized at the outset that these columns could serve as the basis for a book examining what makes special operations "special" and how nations' special operations establishments fit into and reflect their perceptions of their own security requirements and their roles as alliance members. I wish to thank Mr. Copley for his early support and continuous encouragement.

Special Operations and National Purpose incorporates elements of that series as they pertain to the United States, the Soviet Union, and nations of the two European adversary alliances. It also contains a great deal of new or updated material. Where appropriate, I have acknowledged quoted sources in the text itself; however, the vast bulk of the material is the result of my own experience and observations in the course of a decade and a half of association with special operations and the experience and observations of others in the special operations community worldwide. They are too numerous to cite individually even if they desired such recognition, which of course they do not. However, some individuals who have been of particular assistance are due some special thanks. Bob Mountel, an exceptionally thoughtful doctrine and force structure analyst who gives the lie to the "snake eater" special forces image, has been a consistently valuable and cheerful source of comment on U.S. special operations forces. Harry Klein and Linc German, exceptional talents in counterterrorism and special operations command and control issues, respectively, gave of their limited time to review the manuscript. Patricia Gunn helped update the information on U.S. psychological operations equipment plans. Chris Mitchell, a British special operations communications authority, provided information for the chapter on communications. I also drew heavily on the experience of my friend and associate, Ed Tolliver, in developing the section on U.S. low-intensity operations in Central America, and in general discussion. Finally, none of this would have been possible without the patience, good humor, and sheer stamina of Kathy Bublitz, who saw the manuscript through one "final" revision after another. She is a true professional, whether at the word processor or holding a rifle in a mock hostage-barricade exercise, and a valued comrade.

To these and to many others in the international special operations community who have been mentors, colleagues, and friends over the years must go not only my gratitude but much of the credit for such utility as this book achieves. The responsibility for any inaccuracies is, of course, mine alone. It must also be stressed that this manuscript was prepared and submitted while I was a private consultant and is based totally on public record material and personal observations and contacts, and that, notwithstanding my return to government service since that time, views expressed in it are solely my own and do not reflect those of any branch of the U.S. government.

Introduction: The Soldiers of the Shadows

> The Reagan Administration is systematically laying the foundation within the Pentagon for increasing military involvements in Third World conflicts. . . . The new emphasis in many ways recalls the early 1980's, when President John F. Kennedy commissioned the Green Berets to stop what he called the "Communist tide" in poor countries in Asia, Africa, and Latin America. Once again, the Defense Department is beefing up its Green Berets and other "special forces", troops trained to kill guerrillas and to teach other armies to do the same.
>
> —*Washington Post,* June 10, 1984

This excerpt, which begins an otherwise fairly accurate discussion of the status at that time of special operations forces expansion in the United States, illustrates the haziness, bias, and misperception that have characteristically surrounded the subject of special operations, particularly in peacetime. The statement as it stands is not inaccurate, but it is misleading and incomplete: "killing guerrillas and teaching other armies to do the same" is a simplistic rendering of but one of the diverse missions currently assigned to U.S. special operations forces. Most other countries' special operations units are equally subject to one-dimensional public images. Their mission profiles, however, are anything but simple.

The Reagan administration did indeed place significant emphasis on strengthening U.S. special operations capabilities across the board. And certainly many features of that continuing buildup have been reminiscent of Kennedy's enthusiasm for the Green Berets (including the disgruntlement of conventional-minded officers on the service staffs who would rather see special operations–committed resources allocated to additional general conflict–oriented systems). Moreover, while only the Soviet Union has undertaken anything similar recently in the way of a special operations forces (SOF) buildup, most other nations have specially designated special operations units, and many have dedicated considerable resources to their qualitative upgrade in the past few years.

It is perhaps useful at this point to identify what special operations are and, indeed, what makes them "special." There is, as one might expect, diversity of viewpoint. Some analysts exclude counterterrorist operations in friendly territory on the grounds that such operations are in fact police activities. Others define away guerrilla and counterinsurgency warfare, an exclusion that results in special operations' being considered as nothing more than grand raids.

Such views are far too restrictive and in any case do not reflect the thinking of most nations that possess special operations capabilities. For our purposes, a variation on the U.S. Defense Department's definition suffices: Military or paramilitary operations conducted by specially trained, equipped, and organized forces against strategic or tactical targets in pursuit of national military, political, economic, psychological, or internal security objectives. They may support conventional military operations, or they may be prosecuted independently when the use of conventional military or other forces is inappropriate or infeasible. They may include guerrilla and counterguerrilla operations, counterterrorism, intelligence collection, psychological operations, and foreign internal defense training, as well as raid and sabotage operations. They may be conducted during hostilities or in peacetime.

Most certainly, various types of special operations can be and have been carried out by elite troops not specifically designated as special operations forces. The Israeli rescue operation at Entebbe, for example, involved not only the General Reconnaissance Unit but paratroops and members of the Golani Brigade. The airborne and marine forces of the world routinely train for small unit raid operations. Thus, an obvious question is, Why do nations organize and maintain dedicated special operations units?

The answer lies first in why nations have found special operations necessary as national security policy options and second in why many countries have found it advisable to institutionalize their capabilities for carrying out those options.

Special operations in the generic sense address a spectrum of challenges not normally considered appropriate for regular armed military or national forces. Nations facing threats by terrorists and/or insurgents have felt the need to develop specialized capabilities targeted against these threats. Nations with more complex interests and commitments (such as the United States, Great Britain, France, and the Soviet Union) have seen special operations in the guerrilla-counterguerrilla, mobile training team, and strategic reconnaissance roles as offering minimum-escalation options in a variety of politically sensitive situations. In all cases, special operations constitute specific missions or tasks, involving individual risks, to meet specific situations that threaten national interests but that do not warrant commitment of general purpose forces. They may also support conventional military operations, in which case they provide a means of circumventing enemy strengths to attack weaknesses, stimulating popular resistance and chaos in rear areas, and providing otherwise unobtainable intelligence. But they are also available for circumstances requiring military skills in peacetime, when declarations of war are undesirable or when there is concern that introduction of conventional force operations may exacerbate a crisis.

SOF are therefore established and maintained by nations that have identified a requirement to be able to implement a range of specialized military and paramilitary policy options without being forced to resort to the use of conventional units. A paratroop unit may be able to conduct or participate in a raid, but a surgical rescue operation or a counterguerrilla reconnaissance campaign may require other skills and, more important, a different planning and execution mind-set. The thrust of conventional force training, after all, is the achievement of consistent performance of routine tasks to the highest attainable average standard. By contrast, the emphasis in special operations is on directing individual skills to the accomplishment of functions unique to a given mission, generally a high-risk one. Improvisation and independent thinking are essential.

Special operations are not based on sets of exotic operational principles that differ substantially from those of normal combat operations. However, the frames of reference, the national versus tactical orientation of their raisons d'être, and the necessity for psychological self-sufficiency on the part of special operations planners and leaders do constitute points of departure from the conventional operations world. Those aspects, together with the mission goals and constraints applicable to special operations (particularly those carried out in peacetime), require signficantly different priorities in the application of common operational principles. Special operations require sustained independent action and operating in circumstances for which regular force personnel are emotionally and mentally unprepared.

It is to have the capability to conduct special operations on demand, without extensive additional training and orientation, with personnel psychologically attuned to a broad spectrum of unconventional requirements, that nations have developed SOF. Since none of the perceived threats—subversion, terrorism, revolutions, proxy brushfire wars, or conventional military confrontations—that provide the rationale for their existence show any signs of abating as conditions of life in the late twentieth century, it is axiomatic that some degree of policy emphasis on SOF will continue in defense planning throughout both the NATO countries and those of the Warsaw Pact. But policy emphasis is one thing; how it is translated into operational capabilities and doctrines is quite often another. Thus is the direction of inquiry set; how well are various nations truly prepared to carry out operations in the shades of gray? And how do their SOF reflect national perceptions of both individual and collective security roles?

Chapter 1 explores those questions as they pertain to the extensive SOF establishment of the United States and their missions. The Reagan administration's revitalization program will be reviewed, along with major trends in doctrine and hardware.

The second chapter focuses on the European NATO special operations forces. After an initial overview of how alliance versus national considerations affect national SOF establishments and missions, this chapter briefly looks at the forces, doctrines, and salient issues involving the principal NATO SOF elements, including the counterterror forces.

We then turn by way of comparison to the special purpose forces, or *spetsnaz,* of the Soviet Union and their use of both general war-fighting and intervention operations such as Czechoslovakia and Afghanistan. Chapter 4 follows naturally; it addresses the special purpose forces of remaining members of the Warsaw Pact. The focus is how the Soviets have apportioned missions for their allies as much out of belief of potential political danger as of war-fighting utility.

Chapter 5 turns to one of special operations' operating environments, what U.S. doctrine calls low intensity conflict. Here we will compare the Soviet view of extraterritorial counterinsurgency operations with Western views, using recent Soviet support operations as illustrations. Subsequently the multifaceted U.S.-supported counterinsurgencies in Latin America and the long-running British operations in Northern Ireland will be studied.

While the book is about roles of NATO and Warsaw Pact SOF forces, a number of the neutral nations of Europe have recognized a requirement to prepare for both the exigencies of countering terrorism and the unlikelihood of their being able to remain aloof from a general conflict. This recognition has been reflected in SOF development and mission decisions that may have a bearing on overall considerations of European security. The unique challenges to and responses by the neutrals are reviewed in chapter 6.

The next two chapters consider general topics affecting SOF, particularly those in the West. Chapter 7 examines SOF roles in antiterrorism and counterterrorism, establishing among other points that the most obvious role, reaction, is far from the only one. Chapter 8 addresses a deployed special operations unit's lifeline, communications. In most military applications, a deployed SOF element's utility is nil if it cannot report, and the challenge to low-probability-of-intercept, high-speed-data, reliable communications has never been greater. This chapter details how special operations–dedicated communications have evolved in the West and where, particularly in the U.S. and British special operations establishments, they may be going.

The final chapter attempts to place all of this information, and the essential rationale for SOF in the late twentieth century, in a functional perspective. I hope the questions I have set will in large measure have been answered, if not in all cases finally, for there can be no final statement. Where SOF will be, in both NATO and the Warsaw Pact, in the years ahead will depend then as it does now on national and alliance perceptions of need. There is activity now in some Western forces, but a few years ago there was precious little. And in East bloc armies as well as those of the Western democracies, it is axiomatic that any military capability identified as "special" will have, at best, surge priority. Chapter 9, then, offers a few last reflections on how well decision makers have thought through the moral, political, and resource support implications of the capabilities they have lately found so necessary.

Abbreviations

ANGLICO	Air-Naval Gunfire Liaison Company
BBE	Special Assistance Unit—Marines (Netherlands)
BIAT	Antiterrorist Infantry Batallion (El Salvador)
BSG	Bundesgrenzschutz (Federal Republic of Germany)
CA	Civil Affairs
COMSUBIN	Comando Raggruppamento Subacquei ed Incursori (Italy)
CT	counterterrorist
C^3I	command, control, communications, and intelligence
GEK	Gendarmerieeinsatzkommando (Austria)
GIGN	Groupe d'intervention de la gendarmerie nationale (France)
GIS	Gruppi Interventi Speciali (Italy)
GSG-9	Grenzschutzgruppe-9 (Federal Republic of Germany)
IRA	Irish Republican Army
INLA	Irish National Liberation Army
JCS	Joint Chiefs of Staff
JSOA	Joint Special Operations Agency
MEU	Marine Expeditionary Unit
MEU (SOC)	Marine Expeditionary Unit (Special Operations Capable)
MTT	mobile training team

NATO	North Atlantic Treaty Organization
NORAID	Irish Northern Aid Society
NSC	National Security Council
NSWG	Naval Special Warfare Groups
NSWTG	Naval Special Warfare Task Group
NSWTU	Naval Special Warfare Task Unit
NSWU	Naval Special Warfare Unit
OSD	Office of the Secretary of Defense
PRAL	long-range reconnaissance patrol (El Salvador)
PSYOPS	psychological operations
SAS	Special Air Service (Great Britain)
SBS	Special Boat Squadron (Great Britain)
SOCOM	Special Operations Command
SDV	SEAL Delivery Vehicle
SEAL	sea-air-land teams
SOF	special operations forces
TAVR	Territorial Army/Volunteer Reserve

1
The Role of U.S. Special Operations Forces

Revitalization and Reorganization

The U.S. military's special operations forces (SOF)—Green Berets, Sea-Air-Land teams (SEAL), Rangers, and others—after over a decade of post-Vietnam orphan status, have been making a comeback. The Reagan administration found their utility in a muscular foreign policy attractive, as did the original SOF patron, John F. Kennedy, in the early 1960s, and for much the same reason: to thwart Soviet expansion and destabilization activities targeted against the Third World. At the same time, there has been increased emphasis on two other special operations–related areas, counterterrorism and better human intelligence collection. With an accord that is rare in U.S. political affairs, the Reagan administration and Congress saw the need and took steps to refurbish and strengthen individual services' SOF elements and joint service command and control structures.

The policy-driving rationale for prioritizing SOF programs was best articulated by then Secretary of Defense Caspar Weinberger in the Defense Department's fiscal year (FY) 1985 Annual Report to Congress:

> Buttressed by their massive buildup of strategic and conventional forces, the Soviets have undertaken, both directly and through surrogates, a global campaign of destabilization, focused on the Third World, that seeks to obtain their objectives without direct confrontation with the United States. This is, and will continue to be for some time, the most prominent direct threat to U.S. national security interests. . . . U.S. special operations forces are

> being employed to counter these destabilization efforts. . . . By assisting others to prepare their own defenses, we enhance the Free World's ability to cope with Soviet expansionism, reduce the likelihood that U.S. forces will become involved in combat, and demonstrate our determination not to default on our commitments.

He went on to cite SOF roles: unconventional (guerrilla) warfare, counterterrorist operations, security assistance, psychological operations, direct action missions (unilateral raid and sabotage operations), and selected intelligence operations—quite a full plate, and somewhat more involved than simply "killing guerrillas and teaching other armies to do the same." Weinberger concluded his statement on SOF by saying: "Recognizing that the value of these forces transcends the limited investment in resources they represent, we have given high priority to revitalizing Special Operations Forces."

This "revitalization" has been well (some would say overly) publicized, with the attendant consequences; the degree to which force development reality has kept pace with the administration's rhetoric has become the subject of considerable debate within the U.S. special operations community and in Congress. Given overall budgetary constraints and the powerful constituencies within the individual service staffs and on Capitol Hill for more conventional force structures, missions, and weapons programs, it is unsurprising that serious questions have periodically arisen concerning the services' commitment to developing SOF capabilities. It was the questioning of that commitment, in fact, that led Congress in late 1986 to create a unified special operations command and a new assistant secretary of defense for special operations and low intensity-conflict position. However, it is also true that substantial progress has been made since about 1981 in SOF arming, manning, training, deployment, and sustainment and that despite a number of remaining issues regarding training, resource allocation, doctrine, and equipment modernization, the United States possesses the largest, most multidimensional, and most diversely capable SOF establishment in the non-Communist world.

The impetus for the first major event in SOF reorganization, consolidation, and revitalization was actually conceived during the

Carter administration in the blood and flames of the failed Iranian hostage rescue attempt. It was given birth in the Holloway commission report detailing the contributing causes for the disaster at Desert One. Principal among the several recommendations was a call for an institutionalized command and control structure for joint contingency operations, with dedicated forces from the several services. The Carter administration responded, setting in motion the organization of a Joint Special Operations Command, which would serve as a command-and-control headquarters for high-risk overseas contingency operations, and an administration headquarters for such dedicated strike forces as might be assigned.

It was a start, and with the Reagan administration came some policy planners, both civilian and military, prepared to go much further. Noel Koch, then the principal deputy assistant secretary of defense for international security affairs, was "double-hatted" as director, Office of Special Planning. He was given cognizance over SOF revitalization, which he approached single-mindedly and energetically—so much so, in fact, that he was ultimately considered overzealous and, in 1986, asked to step down. Before that happened, however, he was able to make considerable progress on a number of SOF initiatives despite instances of less-than-enthusiastic cooperation from the services and the Joint Staff:

Requiring the services to provide time-phased master plans in which agreed-upon SOF upgrades to meet validated theater mission requirements would receive specified budgetary, manning, and equipment procurement support.

Creation of a Joint Special Operations Agency, headed by a general officer, out of the old Joint Chiefs of Staff (JCS) Special Operations Division, which had been headed by an army colonel.

Creation of special operations commands out of theater unconventional warfare planning cells to upgrade and institutionalize theater commanders' involvement in SOF deployment planning and command and control.

Actual increases in numbers of army Special Forces and Ranger battalions and navy SEAL platoons and special boats.

Apparent progress notwithstanding, however, intraservice and interservice squabbling over SOF asset "ownership," service insistence on budgetary autonomy and prioritization authority, and in particular widely perceived air force mismanagement and lack of support for SOF airlift programs led Koch to become increasingly critical of the services and increasingly insistent on action at the deputy secretary of defense level. When that deputy was Paul Thayer, Koch had a receptive audience; but by mid-1985, William Taft, and his new assistant secretary of defense for international security affairs, Richard Armitage, were disinclined to press the services with the same confrontational enthusiasm that Koch was, and Thayer had been.

With Koch's departure, pressure on the services from the Office of the Secretary of Defense (OSD) for adherence to their master plan goals lessened. However, it increased tremendously from Congress. The late congressman Dan Daniel (D–Virginia), then the chairman of the Readiness Subcommittee of the House Armed Services Committee, had established a special operations panel under Earl Hutto (D–Florida). They found allies in two senators, Sam Nunn (D–Georgia) and William Cohen (R–Maine), already determined to end the harmful effects of interservice rivalry through legislation designed to give greater power to the JCS chairman and the theater commanders. Together they manifested deep skepticism that the services, the Joint Staff, or OSD would, left to themselves, provide the necessary push for seeing the master plan improvements through or ensuring and protecting consolidated SOF fundings, training, career development, doctrine, and operational employment. Accordingly, over OSD and JCS objections, they drafted and succeeded in carrying legislation through Congress mandating an all-service Special Operations Command, an assistant secretary of defense for special operations and low-intensity conflict, and a low-intensity conflict board at the National Security Council (NSC).

A number of issues have not yet been resolved, new questions have been raised, and Congress has been unhappy with how the Department of Defense implemented the legislation. Those aspects, as well as some of the fundamental disputes, will be detailed in due course. At this point, however, it is appropriate to look at the forces themselves.

Forces, Missions, and Capabilities

The Forces

Army Special Operations Forces. The U.S. Army's special operations forces consist of Special Forces (Green Berets), Ranger, Psychological Operations (PSYOPS), and Civil Affairs (CA) units. Those in the continental United States are grouped into the 1st Special Operations Command (1st SOCOM), headquartered at Fort Bragg, North Carolina. The major units and their deployments are:

5th Special Forces Group, Fort Bragg, North Carolina, and Fort Campbell, Kentucky.

7th Special Forces Group, Fort Bragg, North Carolina; one battalion in Panama under operational command of Southern Command.

10th Special Forces Group, Fort Devens, Massachusetts; one battalion in Germany under operational command of European Command.

1st Special Forces Group, Fort Lewis, Washington; one battalion in Okinawa under operational command of Pacific Command.

4th Psychological Operations Group, Fort Bragg, North Carolina.

96th Civil Affairs Battalion, Fort Bragg, North Carolina.

1st Battalion, 75th Ranger Regiment, Fort Lewis, Washington.

2nd Battalion, 75th Ranger Regiment, Fort Stewart, Georgia.

3rd Battalion and Headquarters, 75th Ranger Regiment, Fort Benning, Georgia.

There are additionally a Special Forces detachment in Korea, several other small Special Forces elements in Europe, and a dedicated special operations helicopter unit headquartered at Fort Campbell, Kentucky.

The army also has four reserve components Special Forces groups. Two—the 11th and 12th—are army reserve groups, and the other two—the 19th and 20th—are National Guard units. The reserve components are where the preponderance of the "Total Army's" PSYOPS and CA units are located. Two PSYOPS groups, a CA brigade, and a host of independent battalions and companies back up the more limited active duty capabilities.

The reactivation in 1982 of the 1st Special Forces Group (it was the last of the post-Vietnam special forces cuts) was the first major step in army SOF revitalization. The third Ranger battalion and the Ranger regimental headquarters were also activated in early installments of the army's SOF master plan.

Yet another Special Forces group, the 3rd, is on the books for actiation by about 1990, and the 1st Special Operations Command has developed a mobile signals element to give the command a field operations support capability. Also filling out are the additional subelements for the aviation component, though progression of that unit to full realization of intended capabilities will likely fall victim to army budget cutbacks.

This seemingly bright force enhancement picture is somewhat clouded by the realities of manning the new units. The personnel to accomplish these increases in SOF are not coming from any dramatic increase in army active duty end strength but rather from within the existing army ceiling through the personnel "savings" realized in converting several active army divisions to light divisions, the likelihood that some of the new SOF units (as well as the light divisions) will not actually attain their authorized manning levels, and such other personnel actions as the freeing up of over 5,000 Signal Corps spaces through introduction of automated mobile subscriber equipment.

So while the army has made progress in fleshing out the new 1st Special Forces Group and the 3rd Ranger Battalion, increased authorizations for SOF force structures are still difficult to match with increases in actual personnel assignment allocations. In some cases, in fact, they have gone down, so that Special Forces "A" teams (the basic twelve-man operational unit) are sometimes as much as three to four men understrength. Limiting the army's manpower ceiling may be politically popular, but it renders figures of X number of Special Forces teams illusory. In one extreme case, a unit actually expanded its paper strength in numbers of teams while projecting a decline in anticipated personnel loss versus gain ratio.

Navy Special Operations Forces. The navy's SOF are formed principally into two multifunctional Naval Special Warfare Groups (NSWGs). NSWG 1 is located in Coronado, California, and NSWG 2 operates out of Little Creek, Virginia. Each group consists of SEAL teams, special boat squadrons, and SEAL delivery vehicle (SDV) teams. NSWG 1 has three SEAL teams: the 1st, 3rd, and 5th; NSWG 2 has two that it owns, the 2nd and 4th, plus administrative control over SEAL team 6, the navy's counterterror element. Additionally, Naval Special Warfare Units (NSWUs) with SEAL and special boat detachments, are located in Scotland, Puerto Rico, and the Philippines. The navy has been directed to place its NSWGs under the command of the navy component of the new joint U.S. Special Operations Command; the navy wishes to maintain the groups under the control of the Atlantic and Pacific Fleet commanders.

In the navy, reserve SEAL and special boat squadrons make up a portion of an active NSWG's total strength, much as round-out National Guard brigades do for several active army divisions. There are, of course, advantages to the round-out concept; since most of the SEAL reservists tend to be clustered near an active group's location, they are able to train with their active counterparts on a reasonably regular basis. However, the extremely low active-to-reserve SEAL turnover (due in part to commendably high retention on the active side), currently the navy's only means of manning its reserve SEAL units, has left many unfilled slots in reserve SEAL ranks, particularly in the enlisted billets.

Accordingly, the Department of Defense has included an increase of active SEAL platoons (the sixteen-man basic SEAL element) in the SOF revitalization program. By 1991, there should be seventy SEAL platoons, up from the current forty-one. In 1982 there were fewer than twenty platoons; the "increase" came about as a result of merging the old underwater demolitions teams with the SEALs. Moreover, the navy is bringing a new family of multimission patrol boats into the inventory and meeting its delivery plans for SEAFOX surface craft and submersible SDVs.

In support of SEAL operations, though not directly subordinated to the NSWGs, are one reserve light helicopter attack squadron on each coast. These squadrons train often with SEAL elements and are skilled in off-carrier and night flight operations.

Air Force Special Operations Forces. The air force created a new unit, the 23rd Air Force, to serve as an umbrella command for its special operations elements worldwide. These currently consist of six MC-130 Combat Talon aircraft, ten AC-130 Spectre gunships, nine HH-53 Pave Low, and six UH-1 helicopters grouped into the 1st Special Operations Wing at Hurlburt Field, Florida; four MC-130s each in special operations squadrons (along with other aircraft) in Germany and the Philippines; and four more helicopters in Panama. These are all in active air force units.

The air force reserve has another ten Spectre gunships, and the Pennsylvania Air National Guard has four EC-130 aircraft utilized for aerial PSYOPS missions. Also assigned to the special operations aircraft organizations are combat control teams, which parachute into denied areas either alone or with army special forces elements to survey potential drop zones and landing fields and control airflow.

The Combat Talon and the HH-53 Pave Low are used in long-range infiltration, exfiltration, and rescue operations. There are too few of them, and they are aging. The air force has received only one new Combat Talon in the past four years. As part of its plan for recognizing the emphasis being placed by the administration on SOF, the air force originally proposed to buy only ten more Combat Talons. The Department of Defense told the air force to double that figure.

The air force also had the task of developing a replacement for the HH-53s. Initial air force limits on funding led to greatly reduced capabilities of the prototype HH-60D Nighthawk. This program was to have been given to the army in exchange for the Patriot air defense missile system, but the entire SOF aircraft mission question became an issue of major proportions defying sensible resolution. Thus, the army set about increasing its own capabilities. Finally, in late 1986 and 1987, the air force turned dramatically around, funding not only new MC-130s but improved Spectre gunships, production of significant quantities of the CV-22 Osprey, and new helicopter orders (the RH-53 variant of the "Jolly Green Giant").

Counterterror Operations. Dedicated multiservice forces specifically trained for hostage rescue and other counterterrorist missions have greatly expanded since the ill-fated rescue attempt in Iran. While specific capabilities are classified, they address ground and maritime threats with dedicated units from each of the services. The units train regularly with other federal counterterror forces and allied counterparts.

Of perhaps more importance is the collection of elements into a single whole, commanded by an institutionalized headquarters with extensive operational and intelligence links. This step, born out of recognition that the ad hoc command arrangements prevailing at Desert One contributed to mission failure, has provided consolidated training and mission planning management, intelligence tasking and analysis, and multiservice doctrinal development.

In addition, some theater commanders have tasked their own army special forces and marine corps force reconnaissance units with a contingency counterterrorist (CT) or CT-support role, and the marines recently activated two antiterror security battalions.

Special Operations Organizations. The services, the unified and specified commands worldwide, and the JCS have on their own or under congressional mandate made much progress in centralizing special operations command and control, unit training management, doctrinal development, and operational planning. The army took the lead in 1982 in developing its 1st Special Operations Command out of a previous special operations parent organization, the JFK Center for Military Assistance. (All of the individual special forces, PSYOPS, CA, international studies, and military assistance training courses and schools have been split out to a resuscitated JFK Special Warfare Center.)

The air force followed suit in creating the 23rd Air Force. The navy for some time was the sole service exception. With its two major special operations units assigned to the Atlantic and Pacific fleets, the navy had no centralized parent SOF organization but rather managed training and doctrinal issues from a dedicated section of the navy staff in the Pentagon, through the major subordinate commands, to the units. However, in late 1986 and early 1987, the navy created a special warfare center to centralize doctrine and training and a small special warfare command to interface with the new U.S. Special Operations Command on navy SOF deployment.

The Marine Corps has become involved in the overall strengthening of special operations capabilities too. The marines continue to point out that they have no dedicated SOF per se, though the skills and scope of operations of the force reconnaissance companies offer abundant ammunition for arguments to the contrary. However, the corps has taken the position that certain types of special operations have always been marine missions and that its basic air-ground task force cadre, the Marine Expeditionary Unit (MEU) headquarters, could, with additional training and some special unit attachments, carry out those and more. Thus was born the concept of the special-operations-capable MEU, or MEU(SOC). Contrary to the views of some skeptics, though, the marines have tried to make MEU(SOC) more than just a designation change, designed to get the marines on the SOF bandwagon without having to create new or give up any existing forces. The candidate MEU undergoes some special operations awareness and joint service doctrine training and a validation exercise before being given the SOC sobriquet. And when fielded, the MEU(SOC) now has directly attached SEAL, force recon, and ANGLICO (Air-Naval Gunfire Liaison Company) elements to augment its regular battalion's and air group's capabilities.

Most of the unified and specified theater commands have had staff elements responsible for special operations planning for some time. They have now been reorganized as cadre special operations commands. This peacetime planning entity can be augmented with active and reserve personnel during a crisis or in wartime to provide effective theater command-and-control organs for such SOF units as are assigned.

At the JCS level, the Joint Special Operations Agency (JSOA) was activated on January 1, 1984. JSOA was intended to provide Joint Staff cognizance over, and advise the chairman of the JCS in, all special operations and related activities, including strategic planning, resource programming and allocation, joint doctrine, exercise and readiness evaluation, employment of forces, and interoperability with allied special operations and doctrines. JSOA was to also serve as DOD's focal point for interagency counterterrorism planning and provide Joint Staff supervision over multiservice counterterror forces. However, JSOA's success (or lack of it) in fulfilling its charter as both its creators and congressional supporters had hoped provided much of the rationale for Congress's rejection of JCS assurances that it would take care of SOF on its own, and passage of the legislation creating the U.S. Special Operations Command.

The U.S. Special Operations Command, activated April 23, 1987, is located at MacDill Air Force Base, Florida, somewhat removed from its supervising assistant secretary of defense. Spaces to "pay" for its creation came from terminating the U.S. Readiness Command, also located at MacDill, and JSOA. It has assigned to it the army's 1st SOCOM and the JFK Special Warfare Center at Fort Bragg; the air force's 23rd Air Force, and the special operations training establishment at Hurlburt Field, Florida; the navy's new special warfare center and special warfare command; and the Joint Special Operations Command. Its mandate replicates JSOA's original charter but in addition includes actual SOF command and control in peacetime; selected contingency command and control; joint training and doctrine authority; and, possibly most important, provision for budgetary authority.

Missions and Capabilities

The principal army special operations elements are the Special Forces groups and the Ranger battalions.

Special Forces are charged with training and assisting friendly resistance forces; training and assisting friendly counterinsurgent forces; conducting basic tactics and weapons training for friendly armies; carrying out pinpoint sabotage operations, rescue and recovery missions, and "snatch" raids; and conducting strategic reconnaissance operations attendant to general conflicts and interdictions.

Rangers provide a large-scale strike capability, as well as an additional rescue potential; both dimensions were demonstrated in Grenada. Rangers are organized and trained as elite light infantry battalions, in a modified organizational structure familiar to most soldiers. Where they differ is in their readiness, level of training and motivation, "shock" mission, and the Ranger qualification prerequisite for its leader personnel.

Special Forces, with their more varied missions, perform either as self-contained intelligence, sabotage, or raid elements or as cadres for training and advising indigenous personnel overseas. As such, they require training not only in employing weapons, guerrilla and conventional tactics, and supporting arms but also in the language and instructional skills necessary to teach foreign pupils of varying levels of sophistication. They also need solid grounding in revolutionary and counterrevolutionary warfare principles, intelligence collection, evasion and escape, and special demolitions.

The challenge to Special Forces unit training managers is obvious, and it is one that is both partially met and somewhat compounded by the nearly constant demand for Special Forces mobile training teams. For example, the Special Forces troopers based in Panama have had disproportionately high levels of experience in foreign internal defense training (counterinsurgency, with an ample measure of PSYOPS) of late in El Salvador and Honduras. However, there has been scant time for those units to train in other Special Forces tasks. The same is true in lesser degrees for the other Special Forces groups; nonetheless, the similarity of mobile training team missions to much of the classic wartime requirements of Special Forces units, to say nothing of providing a real-world showcase for Special Forces capabilities, is perceived as a counterweight to areas of adverse training impact.

PSYOPS and CA units are intended to preempt and co-opt adversary motivational themes and to reinforce a friendly power's (or temporarily, U.S. force's) civil administration capabilities. They are organized primarily to support conventional task force and theater commanders but work closely with Special Forces in most of the latter's roles, particularly the foreign internal defense operations.

All army SOF suffer in some degree from a common malady, one for which the ongoing DOD revitalization program provides only a partial cure: the authorized manning levels, operational missions, scope of training requirements, and limited training funds and resources cannot help, when placed against each other, but result in serious capabilities shortfalls. Even were the operational detachments (or "A Teams") fully manned and fully trained in their specialties, the number of teams available, active and reserve, still fall short of worldwide validated wartime requirements by a significant margin.

The reactivation of the 1st Special Forces Group and planned reactivation of the 3rd does make inroads on the total force requirement deficiency, but it also exacerbates to some degree the individual manning and specialty training problems. Active duty PSYOPS and CA units face similar training and manning challenges.

Some of the same problems are true of the navy SEAL elements.

The extensive training required for full SEAL qualification and individual specialties makes manning the additional authorized SEAL platoons as great a challenge as the fleshing out of army SOF. SEALs conduct harbor defense and obstacle reconnaissance, amphibious sabotage, and underwater demolitions. They can conduct limited guerrilla warfare operations and provide maritime patrol and commando training in foreign internal defense programs. The SEAL platoon shortfall is not quantitatively as serious as the Special Forces Operations Detachment deficiency and the planned increases will help further, but expanding the training and skills base is another matter.

Air force special operations aim mainly at long-range insertion into, and extraction or rescue of personnel from, denied areas. For these missions, air force SOF use MC-130 Combat Talons, HH-53 Pave Low, UH-1 and UH-60 aircraft, and Special Operations Low Level (SOLL) crews flying conventional MAC C-130s. Equally important (and warmly remembered by those of us fortunate enough to have had them available in combat) are the AC-130 Spectre gunships, which can deliver automatically sighted 20 mm, 40 mm, and 105 mm fire with exceptional accuracy. These are used for ground special operations support and in support of other joint task force elements in a low- to mid-intensity conflict environment. Other special operations aircraft include the EC-130s of the Pennsylvania Air National Guard, which provide a unique airborne television and radio broadcasting capability in support of PSYOPS.

The significant shortage of long-range insertion and extraction aircraft greatly limits the speed with which Special Forces units could be delivered into denied areas. In wartime, many theater special operations commanders would be forced to choose between scrapping their planned infiltration schedules or dangerously lowering survivability of infiltrating elements by relying on conventional C-130s.

Obviously, isolated operations could be supported out of the current special operations aircraft inventory, but not a full-scale theater unconventional warfare campaign. Defense Department recognition of air force neglect in this area led to the pointed demands to get more MC-130s. In the interim, the air force is also increasing the number of MAC C-130s and crews designated as SOLL, and has embarked on procurement of the CV-22 Osprey tilt-rotor aircraft, the RH-53, and HH-53D Pave Low III.

Performance

This is not a campaign account or a unit history. U.S. special operations in World War II, Korea, Vietnam, Desert One, Grenada, and elsewhere have had and will continue to have their own chroniclers. Where recent or current special operations illustrate points about their utility to U.S. policies and conventional military operations or their forces' command-and-control challenges, some operations analysis is appropriate. That said, the most extensive and ongoing U.S. special operations efforts in the field currently involve antiterrorism and counterterrorism and the low-intensity conflict campaign being waged in Central America. They are detailed in chapters 5 and 7 and addressed elsewhere as well, so further comment here would be redundant. It may, however, be useful in the general consideration of SOF missions, especially since the commencement of the revitalization initiatives, to look at the 1983 Grenada campaign, for here are excellent examples of the actual and potential utility of SOF and their use and misuse by conventional commanders.

Pre-invasion SOF missions included seizure of the Point Salines airfield and elimination of the enemy ground and antiaircraft threat in the area, to permit rapid air landing of follow-on troops and supplies; rescue of American students at the island's medical school; seizure of Radio Grenada, the government's principal means of providing civil and military instructions; and seizure of the governor general's residence and protection of the governor general until relieving forces could move him to a point where he could assert civil control. There were other missions, but these suffice as examples of logical missions for SOF. Moreover, they were assigned to the right elements. The Rangers did take the airfield and move on to the main medical school campus to effect rescue of the students. Radio Grenada was taken by the SEALS, albeit after an initial failure. The governor's residence was taken (also by SEALS) and held for much longer than anticipated, against several Grenadan People's Militia counterattacks. The troops performed exceptionally well. The Ranger commander's decision to drop at 500 feet, under the depression angle of the antiaircraft guns situated on the heights around Point Salines, is smart, bold, independent special operations decision making at its best.

But one SEAL landing came to naught, with several men drowned. When the Rangers liberated one set of students, they found there was another elsewhere. The initial attempt to seize the radio station met with failure. Another grand stroke, an attempt by army SOF elements to carry out a helicopter raid on Richmond Hill prison to free political prisoners who might otherwise be shot, was conceived on the ground and failed. Actions during and subsequent to the drop at Point Salines revealed other planning shortfalls. The common threads in these failures were the "can-do" attitude carried to an extreme in the face of poor intelligence and poor understanding of overall SOF utility in a campaign of this nature by its conventional commanders and planners.

The principal SOF utility could have been, should have been, but was not, in obtaining the one commodity the expeditionary force almost totally lacked: accurate intelligence. The one CIA asset on the island had left and would not return. Locations and strengths of Cuban and People's Militia units were dated. Locations of American students were based on brochures and commercial map symbols. The forces going after Radio Grenada, Governor General Scoon's residence, and Richmond Hill had no idea what they were up against. In all these cases and more, SOF could have provided the same eyes-and-ears assistance to the invasion force as did British SOF in the Falklands. No convincing reasons are found in official after-action reports or verbal postmortems to suggest why making this the first and main SOF priority did not occur. Blessed with an abundance of available amphibious-trained army/navy SOF and marine reconnaissance assets, neither the task force commander nor his staff pressed for full use of these assets prior to committing more extensive forces.

Certainly the SOF commanders on the ground knew the targets they were being sent to attack had not been well intelligenced. Certainly they had some right to express reluctance to proceed. But whereas sustained combat can make leaders a little more wary of bright ideas supported by little or no information, American soldiers generally go into a contingency action with a mixture of fatalism and optimism that accepts as routine that their higher staffs do not know much but that they can prevail anyway. Moreover, in an event of this nature, nobody wanted to miss the "big show" or be thought to lack the fire that characterizes good special operations troopers. In fact, many have since said they had misgivings, but when the word is "go," it's "go." The United States is indeed fortunate to have such men. But its senior commanders are paid to use them more intelligently.

And in many cases, they are. In Latin America, U.S. SOF are the "first team" in a major security assistance effort. Elsewhere in the world, they serve on mobile training teams that provide both the United States and the client states results out of all proportion to the relatively small numbers of SOF involved. SOF also figure largely in each theater's contingency plans; and in September 1987, imaginative use of army special operations helicopters and navy SEALS enabled the U.S. commander in the Persian Gulf to not only force cessation of mine laying by an Iranian naval vessel and abandonment of the ship, but later to capture the ship and crew and demonstrate to the world Iran's perfidy.

But the men who design such plans are not always, or even usually, those who implement them. Utilization of SOF elements for intelligence gathering figured largely in island intervention and rescue plans maintained by the U.S. Atlantic Command in 1983. But those plans were not used at Grenada.

Issues, Challenges, and Trends

We have looked thus far at the current and projected force structures of U.S. SOF and touched briefly on the attitudinal, doctrinal, training, and equipment factors. These factors have been partially addressed by Reagan administration and congressional service initiatives, but they present formidable challenges to attempts at final resolution, as most Department of Defense policy-level civilian and military planners are well aware.

Doctrine

The sheer diversity of SOF missions affects all decision making about force structure, training, equipment, communications, and other elements. First, all SOF elements must be able to operate in a variety of climates, cultures, and terrain. Second, while air force, Ranger, PSYOPS/CA, and SEAL missions do not normally require as many distinct doctrinal orientations as do army Special Forces, supporting a variance of theater SOF employment philosophies has some effect on their assignment of planning directions. By far, however, the major element of the diversity factor is the several distinct operational mission orientations that compete for Special Forces doctrinal development and training priorities.

Many North Atlantic Treaty Organization (NATO) and U.S. European Command planners see dedicated Special Forces assets as theater-directed collectors of essential information in enemy rear areas, providing human eyes and ears at a time when a conventional onslaught may have greatly impaired allied technical collection capabilities. A secondary mission would be sabotage and selected interdiction. Guerrilla warfare would be conducted primarily after the conventional fighting had stabilized by follow-on assets. What this means from a training standpoint is that active component units, which have both home station and annual European exercise opportunities to play guerrilla warfare scenarios, have had to reorient training priorities to the strategic reconnaissance mission, while reserve component units, whose training time and resources seldom permit involved guerrilla warfare scenarios, are seen as the primary practitioners of that end of the special operations spectrum.

Another school of thought holds that, no matter what the mission, the "Tito-in-the-hills" rural-oriented skills are passé. The virtual elimination of virgin, non-state-controlled forestland in Europe, coupled with the sophisticated, multilayered, internal security infrastructure common to our Warsaw Pact adversaries, has in various units at various times led to emphasis on urban guerrilla techniques, which can serve both unilateral and resistance support operations. The success of this technique depends on extensive prehostilities cultivation of potential in-place assets, a condition that Special Forces training includes but no working-level special operations planner assumes.

Still another holds that conventional warfare employment of SOF behind enemy lines will be dramatically attritive, that Special Forces' most telling—and most likely—future contributions will be in the collection of disciplines (counterinsurgency military operations, small unit training, psychological and civil affairs operations, political warfare, intelligence) known as foreign internal defense

(FID). The array of skills required for FID operations are virtually the same as those necessary for organizing and training a guerrilla movement; the difference is the stabilization orientation of security assistance programs versus the destabilization goals of guerrilla warfare. FID proponents believe the only viable means of countering an insurgency against a friendly foreign power is through advice and assistance to internal elements and that if the 82nd Airborne or the marines are required, there has been an escalation that is difficult to justify within the affected country as well as in the United States.

Those who would like to see the emphasis kept on guerrilla warfare skills (which, again, are not dissimilar from those associated with FID) note that most possible adversaries of the United States have potentially or currently restive (and therefore potentially exploitable) minorities comprising a significant proportion of their population, particularly in key border areas.

The problem is that none of these mission considerations can be denied. Strategic reconnaissance, direct action, FID, guerrilla warfare: they all remain generic Special Forces tasks. One theater commander may demand that "his" dedicated active component elements train primarily in one area, to serve his perception of the primary SOF mission, while another will see entirely different requirements and priorities. It is a commander's prerogative—indeed his duty—to establish such priorities, but when those assets are shared between commands or given security assistance missions that have little to do with wartime tasks, the obstacles to intelligent unit-level resource allocation and training planning, to say nothing of coherent doctrinal development, are obvious.

Training

The unique nature of SOF missions presents equally unique training challenges. While an infantry battalion can, with minimal effort, detail off some combat support platoons as opposing forces against which to practice company and battalion-sized operations, SOF require intensive training on a wide variety of individual military, psychological, and language skills; training on imparting that knowledge to foreign regular or irregular combatants; and actual overseas opportunities to put into practice what's been learned.

Moreover, on the C^3I (command, control, communications, and intelligence) and staff preparation levels, running a special operations base overseas is a complex task, one that can only be partially addressed in a simulation environment. One does not develop guerrilla forces or the preparation of an unconventional operational area, complete with intelligence nets, opposition forces, civilian participation, and so on overnight; such training is particularly difficult for the reserve components to organize with their limited time and resources. The heavy commitment lately of SOF elements to Grenada, Latin America, worldwide mobile training teams (MTTs), and overseas exercises has provided a wealth of training opportunities, but it has placed a severe strain on understrength units in maintaining progressive individual or unit training programs. The operations officers and commanders of SOF units over the past several years have performed juggling marvels, but they have been manpower intensive and disruptive marvels; few would claim to have their units staffed and trained to their wartime requirements. Particularly difficult to achieve and maintain, despite its criticality, are language skills. A three-month course, with refreshers, takes up so much training time that not all who need it receive it.

Where unit training is aligned to a given theater's area of emphasis, some skills are being learned very well. However, others (including in some cases what used to be considered "survival" skills related to operating with local nationals in denied areas) suffer as a consequence. Moreover, reassignment of a trooper who has become familiar with one mission area orientation to a unit with a different emphasis has placed both the trooper and his new unit at a disadvantage.

To combat this latter problem, the army is attempting to apply its experimental regimental system (wherein all a soldier's troop duty is spent in U.S. or overseas elements of the same unit) to the Special Forces groups. Thus, a soldier assigned to the 10th Special Forces Group at Fort Devens would spend most of his troop duty time either at Fort Devens or with the 10th's forward-deployed battalion at Bad Toelz, Germany. Attainment of this goal, if achieved, will go far toward minimizing area and language orientation training (and training retention) challenges. The army is also well into a program to build stability on its "A" teams through assignment of warrant officers (who will have served previously as a noncommissioned officer) as team executive officers, replacing the senior first lieutenants previously authorized but seldom available for that position. Finally, the army has made Special Forces a separate career branch, further ensuring assignment homogeneity, particularly among officers.

Another positive factor is an increasing trend toward realism in major unit training exercises. Emergency deployment readiness exercises (little- or no-notice deployments), development of hostile environments in which local law enforcement agencies and other opposition forces actively seek out "friendly" SOF, involvement of (and support for) reserve SOF when possible in the guerrilla warfare scenarios they have previously had scant opportunities to practice: these are all representative of efforts at the unit and theater levels to overcome the impact of externally imposed training constraints.

Command and Control and Force Structure

A number of those who think long and hard about how SOF can fulfill the diversity of assigned tasks believe that the solution lies in task organizing existing Special Forces, Ranger, PSYOPS, and CA units in the army into Special Operations Groups, which would then be supported overseas in the usual fashion by naval and air force special operations elements. They believe existing Special Forces assets should be grouped into specialist elements, such as a strategic reconnaissance battalion, an unconventional warfare battalion, and a strike battalion (which would have a Ranger company as its principal strike element). This concept probably has little future at the unit table of organization level; it not only magnifies unit-level training problems but flies in the face of centralization of Ranger command and doctrinal authority. But it does reflect reality at the theater level. There cadre special operations commands have been created out of operations staff sections that previously oversaw theater special operations planning. These special operations commands—or joint unconventional warfare commands, as they may be known—will, when fully constituted, serve as the theater staff for all SOF forces assigned. These SOF forces may be further assigned directly or in support of a forward operations base with a given functional orientation. Thus, the task-organized special operations group could become a de facto reality in wartime, depending on the theater SOC commander's appreciation of requirements and how he wants to organize his assets to meet them.

The dark cloud on this horizon is that none of the theater special operations staffs has anything like the equipment, communications, or personnel assigned on a full-time basis that they would require to operate on a war footing. They all need augmentation from outside resources. For a limited-scope, low-intensity operation or campaign limited to a single theater, most of such augmentation would come from raiding special operations headquarters and units elsewhere in the active components, with obvious effects on those elements' readiness should they be required suddenly. Here, some service and theater self-help initiatives are being developed in response. One of the theater special operations commands and the army's 1st SOCOM are developing their own mobile communications support capability.

For a general conflict scenario, other problems emerge. In this instance, the major augmentation contingent is composed of reservists. Now, it is axiomatic that the active component units would move to their overseas wartime deployment bases immediately upon initiation of hostilities, and preferably earlier. Yet the establishment of the sustaining, receiving, and deployment infrastructure depends on reservists who must be called up, deployed, and worked into their wartime roles. The services have made some headway against this problem by preidentifying individual mobilization augmentees for theater special operations staffs, and an effort to have such augmentees perform their annual training in their wartime roles and locations is being pursued.

Despite the energy and thought being applied to resolution of organizational issues, SOF command and control continues to remain, in the words of one service SOF planner, the "hottest SOF issue we've got." That it has been such an issue stems from more than just interservice turf battles, which while unsurprising in their existence, have not been (at least at the service staff and unit working levels) quite so virulent as is popularly imagined. It has also been an issue because of residual concern by professional special operations officers and congressional SOF advocates that the services are still not wholeheartedly "with the program." This was, of course, what led to Congress's seizure of the revitalization baton from the Department of Defense in 1986 and the ensuing SOF reorganization legislation.

Doctrine versus Hardware

In any review of the challenges facing the U.S. SOF community and the officials directing the SOF revitalization program, equipment is a major consideration. The uniqueness and diversity of SOF missions had until 1987 caused their associated low-density, highly specialized equipment needs to disappear in individual service budgets. High-profile weapons systems and high-density procurement economies were attractive to congressional constituencies and conventional or nuclear warfare planners alike, whereas SOF found their needs "validated but unfunded." And although the effects of over a decade of neglect are slowly being reversed, nobody involved in SOF operational or resource planning believes that the programmed upgrades cover all areas of equipment shortcomings.

Many of the proposed equipment upgrades represent the results of several compromises with the Department of Defense. Thus even when they are fully implemented, many of the validated force structure support inadequacies will remain. Moreover, in several instances, service implementation of equipment programs already underway has perversely contributed to overall hindrance or deterioration of mission capabilities rather than the desired strengthening.

In special operations, perhaps even more so than in conventional operations, validated missions should drive doctrinal development. Doctrine, in turn, should drive equipment development, procurement, and issue. All too often, the situation in U.S. SOF has been quite the opposite. Limitations on equipment quantity and quality impose limitations on doctrinal practice in both training and in combat. How then, is the revitalization of U.S. SOF dealing with some of the equipment issues?

Airlift

The sophistication of air defense systems throughout most of the regions of the world where U.S SOF would be employed in unconventional warfare, direct action, or strategic reconnaissance roles has long been a source of concern for U.S. Army and joint command SOF planners. Agreeing that the use of aircraft without specialized infiltration-assist Electronic Counter-Measures/Electronic Counter-Counter-Measures (ECM/ECCM) and beacon-homing capabilities for behind-the-lines insertions was an obsolescent technique, the air force almost twenty years ago began searching for alternatives.

The proving ground was Vietnam. Standard C-130s were rigged with various combinations of jamming and detection modules and utilized to support classified cross-border operations. Similarly, the H-3 helicopter, or "Jolly Green Giant" of rescue and recovery fame, was fitted out with ECM equipment. Both were employed successfully in special operations during the war, with the result that standardized production models were introduced: the MC-130 Combat Talon and the HH-53 Pave Low.

With the end of the war, however, also came an end in air force procurement priorities for SOF-related aircraft. Most of the various gun platforms it had developed for special operations and low-intensity conflict support—the A1A and A-37 fighter bombers and the AC-47 and AC-119 gunships—passed into legend and the boneyard. The sole survivor was the AC-130 Spectre gunship.

While the air force did not similarly phase out the SOF delivery platforms, it did curtail procurement. In 1987 only fourteen MC-130s and nine HH-53s were available to meet validated wartime infiltration and exfiltration requirements many times that number.

Perhaps the SOF-related issue that throughout the 1980s has generated the most heat on Capital Hill has been this insufficiency of aircraft designed to move SOF into denied areas safely and dissatisfaction with perceived air force foot dragging and mismanagement of efforts to redress the shortcomings. The issue that prior to the U.S. Special Operation Command (USSOC) enabling legislation received the greatest attention (with least justification) was the on-again, off-again proposed transfer of the long-range SOF helicopter role from the air force to the army. The more important issue was and is the tremendous disparity between validated lift requirements and the numbers of specially designed and equipped aircraft (both rotary wing and fixed wing) either available or projected to meet those needs.

Even with programmed MC-130 upgrades, a significant shortfall in employment aircraft will remain. The current alternative is a standard C-130 with some radar upgrades and specialized crew training—acceptable in low air-defense-threat environments but only marginally above suicidal against more sophisticated protection. Currently twenty additional MC-130s are programmed in the air force SOF revitalization package. Those MC-130s were historically among the first items offered up by the air force when budget cuts were mandated, and management of the entire MC-130 program was criticized in both the House and the Senate. At last, however, seven MC-130s and five AC-130s are in construction.

With regard to the HH-53, the air force was for some time less than enthused with dedication of its rotary-wing assets for special operations other than those associated with pilot rescue and recovery. Accordingly, when tasked with development of new helicopters for army SOF support, the air force originally addressed the responsibility halfheartedly. Funding cuts led to specifications modification; the result, the HH-60D, was transformed from the all-weather, daytime-nighttime insertion-extraction platform it was supposed to be into a fair weather, daytime-only mutant virtually useless for long-range SOF operations, though some modifications and pilot night vision goggle training were carried out.

The army therefore continued to develop its short- and mid-range SOF helicopter support capabilities. Activities included renewed efforts to field an all-weather, day-night-capable version of the Sikorsky Blackhawk, configuration of in-flight refueling schemes for the CH-47 Chinook and conversion of the OH-6 Cayuse observation helicopters to a day-night-capable short-range reconnaissance and firepower platform. These latter were used against the Iranian minelaying ship *Iran Ajr* in September 1987.

To meet the rotary-wing deficiency as well as some of the longer-haul needs, the air force has now committed itself to procurement of the CV-22 tilt-rotor Osprey, which lands and takes off vertically but then flies forward at faster speeds than the HH-53. It has also ordered thirty-three new Pave Lows.

Sealift

Of course, not all means of long-range infiltration are aerial. At one time the navy maintained two diesel submarines with interior compartments modified specifically for the housing and launching of SEAL, marine reconnaissance, and Special Forces elements via Scuba apparatus or rubber boats. After Vietnam, the number dwindled to one, and several years ago that boat too was decommissioned. The navy still accepted the validity of sub-launched SOF missions but expected for some time to perform them with nuclear attack boats. As one might anticipate, the availability of such vessels for training has been highly limited. Many SEALS and virtually all Special Forces troopers have been going through their careers making amphibious insertions from short-range boats or Coast Guard vessels simulating the sub role (the Coast Guard, with its extensive missions and slim resources, has nonetheless proved consistently enthusiastic in its support for amphibious operations trainers of all services). Lately, however, the navy has committed to conversion of old Polaris subs and some diesel boats to not only revive but also expand on the previous capability level.

To support more close-in operations, the navy has procured increased numbers of its high-speed, all-weather SEAFOX and SDVs and has continued its budget commitment for fielding the new patrol boat family in the late 1980s.

Communications

SOF elements deployed in any one of their several behind-the-lines missions have always had specialized communications requirements. First, the equipment must have long-range transmission and receipt capabilities. Second, it must be capable of near-instantaneous burst transmissions of coded messages if it is to defeat enemy direction-finding and decryption capabilities. Third, it must be amenable to air- or waterborne insertion and mobile enough to move quickly. Historically, meeting all three requirements has been elusive for U.S. SOF; with no dedicated procurement element available, SOF have largely had to rely on tried-and-true World War II–vintage equipment or what non-SOF procurers could find. The AN/GRC-109, which was the Special Forces' mainstay for years, was reliable for long distances but bulky, having several components. Moreover, it had no voice capability. The AN/PRC-74, which replaced the 109, was a single-component, man-portable, voice-and-code unit. It could have been the answer to the operational detachments' dreams, but it had reliability problems, drained battery power alarmingly fast, and still could not be netted with conventional forces' radios.

After nearly two decades of work, Special Forces were issued, in 1983, a reliable, multicapability radio, the AN/PRC-70. Its main drawback is that it weighs 45 pounds, hardly making it portable in any SOF operational sense of the word. In addition, it has high battery drainage, so excessive numbers of batteries must be carried. A number of units on their own rejected usage of the AN/PRC-70 in favor of the much lighter HF-only AN/PRC-104, supplementing that with hand-held commercial VHF units for internal communications. Recently the army accepted the 104 as an interim SOF-wide HF radio until another is tested and accepted. Several good candidates exist, including the new British PRC-319, fielded with that country's SOF in 1987–1988.

Army Special Forces and naval special warfare groups have, in recent years, utilized satellite communications equipment and found it extremely reliable, especially in those dead zones too far for tactical FM base-to-detachment communications and too close for "sky-wave" transmissions. Potential problems here are obvious. In a general war situation, satellite survivability is doubtful, and in any case, channel preemption by multiple mobilized theater headquarters is all too likely.

An even more fundamental concern in the communications area is the insufficiency in quantity, particularly in base radios. Special forces groups are supposed to be able to field a special forces operations base and multiple forward operations bases, and the requirements on the group signal companies' tables of organization and equipment reflect that mission. Their authorization level, however, has prevented them from fully supporting that mission or maintaining sufficient stockage to sustain even those base organizations that can be fielded.

Rare, in fact, is the Special Forces group that can consistently field the communications gear required for simultaneous deployment of anything beyond the Special Forces operations base and one forward operations base. Similarly, a naval special warfare group is tasked with creation not only of a naval special warfare task group (NSWTG) but subordinate task units (NSWTUs). In 1987, a deployed group could operate and maintain an NSWTG and one NSWTU only with difficulty.

Thus, as SOF planners consider qualitative communications upgrades, they must also press for quantitative improvement. Parts and components resupply is a problem throughout the services but even more so in SOF, where the low-density, specialized nature of their communications equipment coupled with low authorized stockage levels magnifies the impact of any equipment losses.

Reliable, long-hand communications are central to the effectiveness of any SOF activity but particularly to those forces with strategic versus tactical missions. Some of the measures underway in the United States and the United Kingdom to meet the challenge are detailed in chapter 8.

Weapons

Although specialized weaponry for SOF is usually considered only in counterterror or raid frames of reference (and the counterterror forces are well supplied with state-of-the-art weaponry), other needs exist. For example, while Special Forces and SEAL elements are required to know (and teach others) how to use mortars, few mortars are available for training and none are assigned at the unit level. Similarly, while both wartime operational requirements and many MTT missions include use and/or training others to use TOW antitank missiles, Stinger antiaircraft missiles, and their Soviet counterparts, no tracking simulation equipment (let alone actual missiles) is assigned. Special Forces units at Fort Campbell, Fort Bragg, and Fort Lewis can train with 101st Airborne, 82nd Airborne, and 9th Infantry Division equipment and trainers, respectively, but other units have much more difficulty. Inclusion of a few of each of the above systems in Special Forces and naval special warfare group inventories for training purposes would greatly increase overall capabilities and training standardization.

Air force SOF weaponry, as represented in the 20 mm, 40 mm, and 105 mm systems aboard the AC-130 Spectre gunship, are sufficient in guns-per-airframe terms, but the systems themselves are aging. Moreover, with only twenty of these powerful systems available between active and reserve components, the air force cannot meet the majority of its validated worldwide theater support requirements. Fortunately, the air force recently began a program to convert new C-130s to AC-130 (U)s, a dramatically upgraded version of the tremendous firepower platform. The AC-130 operates at altitudes too high for most tactical air defense systems to be guaranteed a hit; Spectres were, in fact, ever-present lifesavers throughout the six-week-long battle of An Loc in Vietnam, and none was lost, though several other lower-flying aircraft were. They were also used over Grenada. Thus, their use in support of light units in a moderate air defense threat is a proven tactic.

Many SOF units have successfully undertaken to improve their own mission-area survivability by developing sufficient stores of weapons and other equipment used in that area to equip most individual troopers. This is by no means universal throughout U.S. SOF, but it is spreading rapidly. It is yet another salutary example of SOF self-help initiative.

PSYOPS

Some of the once-endemic PSYOPS equipment deficiencies remain, but overall significant improvements have been made since 1986. For example, the heavy printing presses of the 4th PSYOPS Group at Fort Bragg produce state-of-the-art posters and leaflets but are nondeployable. However, progress is being made in what has been a lengthy struggle to obtain procurement priority for a modular print system. When fielded, this system will provide better flexibility and mobility to PSYOPS printing support capabilities.

The army is also aggressively pursuing a number of other PSYOPS equipment initiatives: a 155 mm leaflet artillery round to accommodate the transition of the field artillery from 105 mm howitzer to the lightweight, larger-caliber gun, a graphics production vehicle, electronic optical translator, and improved loudspeaker systems. How many of these programs will obtain and retain budgetary commitment long enough to bear fruit is anyone's guess, but PSYOPS planners are optimistic, which they have not had cause to be in many years.

Other Areas

A great many improvements to SOF parachute delivery, weapons, target location and designation, position location, and transportation are in development or testing. Increasingly the budget dollars are going toward off-the-shelf, nondevelopmental issue items for which the R&D has already been accomplished by another agency, service, or country rather than time-consuming, expensive, new R&D. SOF budget dollars are neither as extensive nor protected as administration assurances suggest; the Program 11 (DOD's SOF budget line item) funds are still administered by the services; and competition within services for overall priorities is, if anything, more intense than ever. There is very much a sense that the SOF community is racing against time, that the funds will soon dry up, and that if it engages in protracted R&D, it will likely realize no ultimate benefit.

The Future: Commitment or Cosmetics?

The expansion and upgrading of U.S. SOF, whether due to service, Defense Department, or congressional urging, has been a direct function of perceptions of SOF utility in counterterrorism, counterinsurgency, and, under certain conditions, aiding anti-Communist resistance forces. Certainly SOF have had a significant role in the Reagan administration's posture toward state-sponsored terrorism and the spread of Soviet-inspired political destabilization. But the continued improvement of SOF capabilities must depend on more than a given administration's Third World–oriented security assistance enthusiasm. It is unlikely, for example, that a Democratic administration will perceive the regional conflict in Central America in quite the same activist frames of reference as Reagan's.

Whether it does or not, though, or whether a successor Republican administration has the same orientation, SOF capabilities enhancement retains a policy-independent rationale. Those capabilities, tools for use in a dangerous world, represent an ability to choose and execute a wide variety of security assistance and low-profile force projection options short of overt crisis intervention. At the same time, SOF utility in a regional or global general conflict has been accepted. Tools and options for policymakers: these SOF can provide any U.S. administration, regardless of its political philosophy and foreign policy proclivities. It is true that the impetus for SOF revitalization grew out of the Reagan administration's foreign policy priorities, and there is danger there. SOF enhancement has, as the June 10, 1984 *Washington Post* article so clearly indicates, become viewed too narrowly as a function of that policy in some circles.

Certainly there is reason to believe that the longer, non-policy-dependent view of SOF utility, and thus, sustainment of the revitalization momentum, will prevail. After all, the impetus for the most recent, and sweeping, burst of SOF-related initiatives came not from the Reagan administration but rather over its objections from a bipartisan group of legislators in a Democrat-controlled Congress. Will this constituency last? Will a successor administration recognize that robust SOF capabilities represent tools and options, or will it see them as tainted manifestations of particular foreign policies for which it has no enthusiasm? There are certainly grounds for the former, hopeful, view. But the course of America's political history, and its penchant for confusing cause and effect in international security affairs, lend equal weight to a less optimistic reading of the tea leaves. As with all other things, only time will tell.

2
The Role of Special Operations Forces in Western Europe

Alliance versus National Perspectives

The SOF of NATO countries reflect, in their doctrines, training, missions, and resource priorities, the diversity of their nations' threat perspectives, strategic role perceptions, and national foreign policy goals. While all Alliance powers have either directly committed specific units to NATO missions or at least to wartime support of their NATO-committed conventional forces, many units also have nationally driven counterterror, counterinsurgency, or security assistance missions. These can have as much or more impact on those units' force and doctrine development, procurement, and training as do the Alliance-related requirements. Moreover, most European Alliance-related SOF utilization planning reflects constraints imposed not only by overall force structure limitations but also (in the European view) by the operational scope limitations inherent in the essentially defensive NATO missions. This Mulligan's stew of varying orientations and capabilities presents obvious challenges to multinational SOF utilization war planning. To a greater degree than one might suppose, however, these differences have been assimilated and, in many areas, particular national SOF strengths have served the overall planning effort.

European Theater Support

For the most part, NATO countries have viewed the proper wartime roles of their SOF to be a combination of long-range reconnaissance, commando, and stay-behind sabotage operations. Some nations' forces, like those of Italy and France, perceive a requirement to support guerrillas; however, these operations would be conducted on enemy-held friendly territory. Only the United States and, to a lesser extent, Great Britain envisage operations in enemy home territory during a NATO conflict. Part of the rationale for these operational-scope limitations lies with the fact that most NATO countries have neither the overseas national force projection requirements, the perceived economic wherewithal, nor the domestic political constituency for maintaining more sizable active military establishments than those now in place. The impact of such constraints on SOF can well be imagined. As a European special operations expert once explained: "Given that our armies are small, the resources and manpower that will be devoted to special operations will also be small, to the point where even if the proportional amount of our budgets devoted to special operations equaled that of the United States, the realized capabilities in quantitative terms would be quite slim." If the United States has trouble finding the funds for aircraft to be utilized primarily for special operations units' missions, for example, one can visualize without difficulty how the national assemblies of countries like the Netherlands or Belgium might receive a request for specially outfitted aircraft to support their small SOF elements in long-range infiltration missions.

Additional rationale for the European interpretation of NATO SOF missions is drawn from collective European interpretations of the mission (and consequent strategic goals) of the Alliance itself. There are significant differences between Europe and U.S. approaches to war fighting. Colonel Arie van der Vlis of the Army of the Netherlands summarized them succinctly in the Summer 1984 edition of *Parameters,* the journal of the U.S. Army War College:

> You might say that NATO's strategic objectives in wartime are more limited than the objectives that can be derived from FM 100–5 [the U.S. Army's manual on battle operations]. If I understand NATO's strategy well, NATO thinks more in terms of winning battles than on winning a war, in defeating elements of the opposing forces rather than defeating the enemy nations themselves, and trying to restore as quickly as possible the situation as it was, rather than in pursuing objectives beyond that aim. . . . In geographical terms there will be great political pressure

> not to go beyond the objectives of restoring the integrity of NATO's territory. To summarize, NATO does not seek a military victory over the USSR.

These differences provide fuel for a good deal of polite debate between U.S. Army conventional force commanders and their European counterparts. They also have had considerable impact on national SOF mission definitions.

Those definitions, however, have largely been turned to advantage by both national and NATO operational planners. For example, if it is true that only the United States believes in "winning," then only the United States might be expected to plan seriously for special operations in enemy countries. However, as it happens, only the United States has the force structure, budget, delivery platforms, and so on to support such an approach in the first place. Thus special operations in enemy territory would fall largely to the United States (with limited amphibious support in the Baltic and Mediterranean from other countries) regardless of differences in strategic goals interpretation.

By the same token, there is a need for long-range patrol and raiding capabilities to support both the national corps and allied regional commands. The national SOF of Europe are better suited than U.S. SOF for the more close-in missions, and their use in these roles frees the Americans for deeper, more strategically oriented missions. At the same time, the United States too has recognized the need for long-range patrol capabilities to support its corps commanders and has activated two companies for that purpose in Europe. U.S. doctrine considers such functions tactical, as opposed to strategic, in nature and therefore not a special operations function per se. These companies, then, despite the similarity of their mission with those of European SOF, are not considered by the U.S. Army as SOF (though the members of the units in question might be inclined to argue quite forcefully to the contrary). Again, only the United States is in any position to be able to make such distinctions; European SOF must be jacks-of-all-trades.

Thus, the force structure and orientation differences in reality allow for each nation to apply its own strengths to different roles, all of which must be played. Moreover, in a number of countries, there are shared missions—principally the stay-behind sabotage and guerrilla support operations—in which European and U.S. SOF find common cause. In these missions, U.S. and European SOF complement one another's capabilities, and years of joint exercises have developed high levels of interoperability and camaraderie. Such missions are quantitatively decreasing, however, in the peripheral areas where either national SOF capabilities enhancement or mission assumption by another European power has permitted reassignment of the U.S. SOF to other responsibilities.

Power Projection Support

While the colonial empires and wars that necessitated the maintaining of large forces overseas no longer bedevil the European powers, security of their remaining possessions and retained security assistance responsibilities for their former colonies have mandated retention of some power projection capabilities by several nations. SOF have been utilized extensively in both roles. The British and French SOF, in particular, have long had training assistance relationships with both the police and military special operations units of many of their former dependencies. The British utilized the Special Air Service (SAS) and the Special Boat Squadron (SBS) in raiding and reconnaissance operations during the Falklands campaign, and French special operations elements have deployed frequently in intervention operations in francophone Africa. Belgium has had to deploy airborne forces twice to Zaire since granting independence in 1960.

While these deployments have demonstrated the need for retention of special operations (as well as other intervention-related) capabilities for the governments concerned, they have also served to highlight deficiencies. Years of postcolonial orientations to NATO requirements have led to degradation of long-range deployment capabilities in particular, both aerial and amphibious. Both the 1964 (Belgian) and 1978 (Franco-Belgian) parachute rescue operations in Zaire required the use of U.S. long-range troop-lift aircraft. The commando carriers from which Great Britain was able to conduct small-scale intervention or support operations in Kuwait (1961), Brunei (1962), Tanzania (1964), and Cyprus (1974) are largely gone. Before the Falklands campaign, the United Kingdom stood poised to allow its remaining commando carrier and two amphibious assault ships to pass from active service. The latter have been retained, but no plans have been made to replace them at the end of their normal operational life.

National versus Alliance Goals

In isolated cases, national missions for special operations forces work antithetically to Alliance goals. Prior to 1974, principal mission orientations for both the Greek and Turkish SOF envisaged typical guerrilla support and stay-behind sabotage operations against Soviet and Warsaw Pact forces. Animosity toward each other was a major but not preeminent aspect of foreign policy. U.S. SOF trained regularly with their counterparts of both countries, as did the British. With the 1974 Turkish occupation of northern Cyprus, the Greek position began to change. It has now been somewhat formalized by Prime Minister Andreas Papandreou, who has declared Turkey to be the principal potential foe, ordered the armed forces to orient their thinking accordingly, and sacked mid- and senior-level officers who cling to outdated Russia-first views. Whether this has any bearing on the technical skill in which both forces are trained is doubtful, and indeed the animosity has perversely led to quantitative increases in manpower and equipment. From a capabilities standpoint, then, the Greek-Turkish tensions may be said to have fostered upgrades that may ultimately be of use to NATO. However, from the standpoints of integrated planning, training, and logistical support, the impact of the standoff is decidedly negative.

Counterterrorism

All NATO countries have developed specialized units to combat terrorist attacks domestically, and some can project that capability overseas. Military units in Britain and the Netherlands have the principal counterterror operations mission but would be used only when paramilitary national law enforcement agency elements are insufficient for the task at hand. And in some cases, such as France and West Germany, custom and law have restricted the counterterrorism role solely to paramilitary national police forces.

Within the military forces, there are some further differences.

Some have counterterrorism as their principal, full-time mission, while others train for it either as one of several ongoing missions requirements or as a mission rotated from unit to unit. Those with harbor facilities and offshore oil platforms (all except Luxembourg) have developed some maritime counterterror capabilities. As might be expected, however, in cases where the lead responsibility is held by national law enforcement agencies, those capabilities are quite limited.

Some rivalry, personality conflicts, and downright jealousy between forces within a given country and between those of the several countries do exist and are natural. However, there is a strong "we're in it together" feeling throughout the counterterror forces of the allies, as well as the professional bond that has long existed between most allied SOF. These contribute to joint training, information sharing, special equipment loans, operational teaming, and other manifestations of mutual support.

A summary of NATO special operations across the board might be stated similarly. National SOF capabilities address national requirements, as well as differing strategic appreciations. Both the strengths and limitations have been factored into NATO planning to optimize the former and minimize the latter.

The United Kingdom

The SOF of the United Kingdom are direct-line descendants of a colorful variety of World War II special operations groups. The two principal elements of the British SOF structure—the army's SAS units and the Royal Marines' SBS—can point to specific namesake antecedents. Their mission profiles also incorporate those once associated with the wartime special operations executive, the Jedburgh teams that parachuted into occupied Europe to aid resistance elements, the regular and ad hoc commando units of the army and the Royal Marines, and the Long Range Desert Group. With but a brief postwar hiatus, British SOF have since seen service in all the nation's "twilight of empire" campaigns, building an extensive institutional training and operational memory in tropical and desert operations. At the same time, the United Kingdom's NATO assignments and the increasingly menacing specter of organized terrorism have led to intensified emphasis on arctic and mountain warfare, long-range and strategic reconnaissance techniques, and counterterror operations.

It is in counterterrorism that British SOF—primarily the SAS—have achieved the most extensive public notice. Prior to the Falklands campaign, the popularity accorded the SAS as a result of its successful recapture of the Iranian embassy from terrorists in May 1980 was felt by the British special operations community to have been a double-edged sword. On one hand, the utility of a highly trained elite force had become obvious, and the equipment procurement checkbook was open to SOF in a way it had never been previously in the general atmosphere of military budget cutting. On the other, counterterrorism was a mission only recently assigned to the SAS (and other British SOF) elements; the glamour accompanying this additional mission was perceived as overshadowing the importance of the primary unconventional warfare missions not only in the public eye but within the Ministry of Defence. However, the 1982 operations in the South Atlantic, which for British SOF included both commando raids and intelligence-gathering missions critical to the successful outcome of the British effort, helped reorient official (if not popular) attention on the more traditional SOF mission capabilities.

Organization and Traditional Operations

The principal British SOF structure is the Directorate of Special Forces (grown out of the old Army Special Air Service Group), which serves as the umbrella command, administration, and training headquarters for SAS line units and SBS. The SAS includes one regular regiment, the 22nd, based with all training and signal activities at Hereford, and two Territorial Army/Volunteer Reserve (TAVR) regiments, the 21st and 23rd. A regular SAS signals squadron supports 22nd SAS requirements, and a TAVR SAS signals squadron supports the two reserve regiments. Each regiment, regular or TAVR, has four squadrons, each composed around four sixteen-man troops, which in turn usually work operationally in patrol teams of four men. In the regular regiment, some variations exist due to specialization requirements of individual troops (mountain troop, boat troop, and so on). Moreover, the

22nd has assigned to it a fifth squadron composed of reservists, which serves as a pool from which specialists for particular missions or overseas security assistance training teams may be drawn.

The principal SAS mission is one of unilateral special operations in denied areas, including sabotage, raids, intelligence gathering, and counterguerrilla penetration operations. A secondary mission is that of organizing and training friendly resistance forces. In peacetime, another secondary mission, which it does extensively, is provision of specialized security assistance training to friendly nations.

Less publicly recognized is the Royal Marines' SBS. Overall strength and organization of the SBS remain closely held, in large part because the SBS has not had the long track record or recent notoriety of the SAS. Its missions are similar to those associated with U.S. Navy SEALS and the Soviet naval *spetsnaz:* coastal sabotage operations and ground, surface, or underwater reconnaissance of potential landing beaches and enemy coastal facilities. While turf distinctions between SBS missions and deployment means and those of the SAS may be as generally true as they are for their Soviet and U.S. counterparts, like them the SAS and SBS have overlapping capabilities. Thus, the Falklands campaign once found SBS reconnaissance personnel deploying via helicopter on a mission similar to those normally given the SAS, while at the same time SAS personnel were deploying in small boats for a mission one might have thought more appropriate for the SBS.

The SAS and the SBS operated extensively throughout the Falklands campaign, demonstrating not only their utility and physical hardihood but their ability to recognize, adapt to, and exploit changing situations. SAS and SBS teamwork played the leading role in the capture of the South Georgia Islands. After some initial misadventures in which, miraculously, no lives were lost but several reconnaissance missions came to naught, successfully emplaced surveillance teams noted a high degree of disorganization among the 140-man Argentine garrison following the British helicopter attack on the Argentine submarine *Santa Fe.* Rather than wait for the Royal Marine company detailed to land and seize the port of Grytviken to arrive (it was still some 200 miles away), the SAS and SBS officers present held a hurried council of war with supporting naval officers and then mustered every SAS, SBS, and regular Royal Marine trooper available in the small flotilla into a composite company seventy-five men strong. This force took off aboard three helicopters and, under cover from three destroyers, landed near Grytviken. The Argentine garrison surrendered shortly after.

Other highlights of SOF operations during the Falklands campaign included intensive SAS-SBS reconnoitering prior to the main British landings and the SAS raid on the Argentine air base on Pebble Island, just north of West Falkland Island. In the former operation, seven four-man SAS patrols were inserted by helicopter at various points outside suspected concentrations of Argentine troops, while six SBS teams inserted by small boat into San Carlos Bay and other possible landing sites. After the decision was made—based on the SAS and SBS reconnaissance reports, since useful tactical intelligence was available from no other means—to land the main force at San Carlos Bay, small SBS and SAS parties covered the approaches and carried out diversionary and clearing operations to ensure unobstructed landings. All were successful, and indeed, the landings at San Carlos were not met by Argentinians but by SBS patrols that had been on the island for twelve days.

The most spectacular and telling operation conducted by SOF elements prior to the landings at San Carlos was the raid carried out against the Argentine air base on Pebble Island. The aircraft based here posed a serious threat to the landing force; they were not far from San Carlos Bay and therefore would have long and potentially dangerous loiter times over the intended landing area. Preceded by a reconnaissance team that confirmed a large number of Pucara ground-support aircraft and a garrison in excess of a hundred Argentine soldiers, a force of forty-five SAS troopers and a naval gunfire support team landed by helicopter in fierce winds and initiated sabotage operations against the aircraft. At the same time, small arms fire from part of the SAS force and naval gunfire called in by the Royal Marine team crushed an incipient counterattack. The group took off in a Force 9 gale, having destroyed eleven aircraft and leaving the base (and its garrison) in shambles.

Those operations, and other diversionary attacks and reconnaissance operations in the Falklands and on the Argentine mainland itself, contributed to an overall revival of understanding within the British defense establishment of the critical roles played by SOF in support of conventional operations. SOF served as the most reliable—and often the only—eyes and ears for the task force. SOF raiding actions materially reduced enemy capabilities, sapped morale, and diverted or pinned down potential reaction forces at critical times. Both activities dramatized the force multiplier aspect of SOF, and the weather and terrain conditions under which SOF operations were carried out reinforced their utility in and relevance to operations in Britain's second area of NATO responsibility (after North Germany), the Northern Flank.

Counterterrorism

The counterterrorism mission in the United Kingdom is a multifaceted one; contrary to popular belief, it is not solely the province of the SAS. The SAS is charged with providing assault-and-rescue forces during terrorist facility seizures in the United Kingdom proper, as well as operating covertly against the Irish Republican Army (IRA). It discharges this responsibility by rotating the counterterrorist assignment among its regular squadrons, each of which is brought to full strength and given several weeks of close combat and counter-IRA refresher training prior to assumption of its counterterror tour. Remaining in readiness is the designated squadron's mission, but it does not do so merely by continual training. Elements are rotated to Northern Ireland, where they conduct intelligence-gathering and interdiction operations in the border regions. Other elements have recently been detailed to guard high-risk foreign dignitaries visiting the United Kingdom, selected British diplomats overseas, and on occasion the prime minister. This mission expansion has not been accompanied by any appreciable growth in personnel authorizations, so the impact on orderly, forecastable rotation of personnel and training schedules has been felt. An additional distractor is the high demand from developing countries for SAS training and assistance teams.

By contrast, the Royal Marines, which were assigned nuclear submarine security and offshore oil rig recapture missions in the late 1970s, ultimately discarded the unit-rotation concept practiced by the SAS in favor of a dedicated unit. Moreover, the Royal Marines succeeded in acquiring sufficient additional personnel spaces that the existing force structure would not be bled to staff the counterterror force. This force, established in 1980, became known as Comacchio Company. Comacchio is supported by elements of two Royal Navy helicopter squadrons located near its Scotland base and an SBS detachment.

The SAS has participated in a number of the major terrorist seizure resolutions in recent years. Specialists either directly assisted or observed in the Dutch termination of the train-school seizure in May 1977 and the German rescue of hostages on a Lufthansa airliner in Somalia later that year. Similar assistance was provided to the Italians during planning for the rescue of U.S. Brigadier General James Dozier. The SAS's most spectacular hour was on May 5, 1980, when its troopers came through the rear, front, and side of the seized Iranian embassy in London, preceded by a barrage of flash, stun, and gas grenades. They rescued the hostages, killed four of the five terrorists, and lost not a man.

Weapons, Equipment, and Training

Volumes have been written on the rigors of SAS training, and deservedly so. As rites of passage into the continuing physical and mental demands of SAS service, the conditioning drills, forced marches through trackless, barren terrain, and resistance-to-interrogation workovers accomplish their grim purposes. After the initiation phases, extensive training in British and foreign weapons; close combat techniques; mountain, arctic, desert, and tropical warfare; foreign languages; intelligence and surveillance; and amphibious operations awaits the successful candidate. Weapons include a wide variety of rifles, submachine guns, pistols, small mortars, and explosive devices; grenade launchers and Finnish sniper rifles are also used.

British SOF have also incorporated high technology into their techniques. In the Falklands they used thermal imagers to verify the presence of personnel in buildings or fighting positions, and the SAS employed a satellite communications system. The SAS uses infrared night observation equipment and various other surveillance, target acquisition, and sensory devices. Although one Royal Marine officer opined that the major effect of the Falklands campaign was not so much to teach new lessons for SOF as to reinforce old ones, some initiatives emerged in the area of equipment. The realities of wet-cold versus dry-cold combat environment factors accelerated several associated procurement actions, notably Gore-tex clothing and pack coverings. Moreover, while existing radios served British SOF well in the Falklands, particularly because all coordination had to be conducted between separate elements on ships, the lack of burst transmission capability was felt. British SOF have since moved to procure a new radio, the AN-PRC 319, manufactured by the British firm MEL, which addresses that and other SOF-related mission requirements.

In equipment, as with manpower authorizations, personnel selection, and training, the British have opted for quality and consistency of performance over quantity. As a consequence, British SOF force structure has been, and will likely remain, small. While some force expansion to keep pace with expanded counterterror missions has been urged in many quarters (successfully in the case of the Royal Marines), the salutary effects of setting and maintaining standards so high only a chosen few can meet them are undeniable, as SAS and SBS performance in the Falklands attests. Will the mission load stretch the skills and personnel base too thin? The answer lies in the future. For the present, British SOF have met each of several widely varying operational challenges—from London, to Northern Ireland, to the Falklands—with success, and success offers little room for argument.

France

As with the rest of NATO's SOF, those of the Fifth Republic reflect national perceptions of European political and military roles. Unlike most of the other European members of NATO, however, France retains significant overseas interests in the Pacific, the Caribbean, and Latin America, and it maintains considerable influence with its former African and Indian Ocean possessions through extensive military assistance and mutual defense agreements. Internally, France's traditionally volatile politics have long provided work for the most extensive internal security apparatus in non-Communist Europe. Thus the counterterror dimension of French special operations also has its own uniquely French character.

The French therefore recognize requirements for long-range

reconnaissance, stay-behind sabotage, and airborne-airmobile strike operations in the European context; overseas counterinsurgency training, advice, and operations; and a counterterror capability that can operate both in metropolitan France or abroad. With the exception of the counterterrorist function, France has relied principally on its paratroopers to maintain capabilities in these diverse areas, as well as those associated with more conventional combat operations.

Les Paras and Special Operations

Since their inception, France's paratroop formations have had virtually continuous experience with small unit operations in a counterinsurgency–low-intensity conflict environment. Their ranks produced considerable counterinsurgency thought and doctrinal development, the result of contact with revolutionary movements in Indochina and Algeria. *Les paras* have long understood the relationships involved in stability operations, whether those operations called for training native forces (as in most of France's former African colonies), stiffening indigenous forces facing internal or external threats (as in Chad in the 1970s, again in 1983–1984, and at present), crushing an urban guerrilla network (as in Algiers in 1956), or carrying out pinpoint raids (as in Zaire in 1978). So while one at first might question whether regular airborne units can be legitimately considered SOF, closer consideration compels recognition that French airborne forces have had more experience with the various areas that we currently think of as special operations than the named SOF of most other countries. In fact, at times they have been too closely identified with political and revolutionary actions. The techniques (arrest without formal charges, torture, and so forth) employed by the *paras* in winning the battle of Algiers in 1956 led to a tremendous outcry in metropolitan France. Later, disgusted with the Fourth Republic, virtually all of the airborne division then in Algeria supported Charles de Gaulle's establishment of the Fifth. Disillusioned with his abandonment of Algeria, many *paras,* including the 1st Foreign Legion Parachute Regiment, supported the abortive army putsch in 1961. To this day, an element of official distrust remains; the number of years in which an officer can serve in the paratroop units is limited.

Most paratroop units are now assigned to the 11th Airborne Division, located in Tarbes near the Pyrenees Mountains. Within the division, one of the brigades (2nd) has been made the repository for maintaining long-range reconnaissance and guerrilla warfare capabilities. Both of these functional areas should be conceptualized primarily in their European context.

Two of the three general mission areas identified for the new Force d'action rapide, of which the 11th Airborne Division is a part, involve European scenarios. The first is operating beyond the range of the 1st Army (France's principal first-line ground combat elements, located in eastern France and southern West Germany). The second is to react to an enemy breakthrough. In both cases, long-range reconnaissance capabilities would be required. In the second, cadre elements would begin organizing overrun civilian populations into resistance groups. While the doctrine and skills involved in unconventional warfare are not official envisaged for operations beyond French territory and peripheral areas, obviously they can have applicability in other overseas scenarios. The regiments of this brigade, including airborne Foreign Legion and marine infantry regiments, are where one would expect to find military free-fall, specialized demolitions, and other nonstandard capabilities.

All the division's units, however, have responsibilities for the rest of what we might consider special operations missions, particularly counterinsurgency operations and training, and strike operations. At least a third of the division is abroad, either on training missions or maintaining presence (and, not incidentally, order) in Africa or the Indian Ocean. In effect, just like Britain's SAS and U.S. Special Forces, the division's units must be able to act as cadres or self-contained operational elements. The self-contained element, however, is more like a U.S. Army Ranger battalion than a Special Forces or SAS unit.

These multiple capabilities are essential to France's power projection in its sphere of influence. The *paras* are legendary; more than once, the mere presence of a company or two, visibly backing up a pro-French African leader, has brought an incipient coup d'état to an end. Their redeployment to Chad when Libyan forces entered the Chadian civil war brought a halt to the forward advance of victorious Libyan-rebel columns and currently provides the Chadians a stable rear area, a training base, and considerable logistical support.

Creation of the Force d'action rapide reflects French recognition that (as in Chad) some power projection operations require forces capable of greater firepower and sustained combat than the *paras* can contribute operation alone; nonetheless, the *paras* will always be the first and principal low intensity or crisis intervention force.

Probably the best recent candidate for a brief case study of both strengths and weaknesses in French special operations is the drop of the 2nd Foreign Legion Parachute Regiment at Kolwezi in Zaire in May 1978. The operation itself was undertaken to rescue several hundred Europeans who were in Kolwezi when a force of rebels invaded Shaba province from Angola and captured the city.

France and Belgium had agreed to come to Zaire's aid and rescue the captive Europeans, with the Foriegn Legion paratroopers being tapped for the initial combat assault. Virtually everything that could happen to frustrate intelligent planning did. Because no long-haul troop-carrying capability was available from the French Air Force, U.S. Air Force C-141s and a C-5A were required. The number of airframes available and the urgency of the deployment prevented the French from taking any but the most minimal equipment. No information was available, then or later, on enemy force strength, deployments, or capabilities. Upon arrival in Zaire, the French force commander found that the aircraft available for the actual drop were limited to a C-160 Transall and four C-130s (a reduction of one each over the anticipated package). Moreover, since there had been insufficient space for their own parachutes, the French had to familiarize quickly with the American T-10s available in Zaire. It was out of the question to reduce the already small number of personnel in the first wave (381), so the force commander made a decision that required both nerve and confidence in his troops' adaptability: he crammed 80 to 85 personnel into the C-130s (designed to hold 64). This prevented any final equipment check and severely limited the ammunition the mortar section could bring.

Given the dearth of information, the commander had decided to attack the terrain (that is, seize assembly areas), then key points near the city, and finally the city itself. The initial jump took place on May 19 on a poor drop zone. Had the commander not decided on a low jump altitude (600 feet), his on-ground assembly would have been even more difficult than it was. When more aircraft problems prevented dropping of the second wave until the following morning, the commander took advantage of the delay to move the drop zone to an area that better supported his task of clearing Kolwezi.

As it turned out, most of the rebels were on the move out of Kolwezi when the operation was launched; they had heard on May 18 that the Foreign Legion was coming. Even so, internal rebel communications and coordination were apparently so bad that those still looting and abusing Europeans in the town were surprised when the French assaulted them on May 20. The Legionnaires stormed into one building just as forty Europeans were about to be killed by their captors. All the captives survived; none of the rebels did. Heavy fighting continued all night as the French kept the pressure on and prevented any rebel consolidation of forces.

The four companies of paratroopers had been flying, marching, and fighting for four days with virtually no sleep. When the Belgian force arrived on May 21, the French moved out of town and slept for a night. Fighting on subsequent days flared off and on as the regiment saw the remaining rebels out of Zaire, but the principal mission had been accomplished. In the face of one difficulty after another, with no reliable intelligence, and contending with severe physical demands and lack of sleep, the *paras'* training and discipline paid off in adaptability, stamina, and morale.

Any force can look good when everything goes as planned and all the envisaged support is available. The true test of a unit is its performance when initial plans go awry, and expected support does not materialize. The traits displayed by the 2nd Foreign Legion Parachute Regiment exemplify the emphasis placed on independent action, flexibility, and improvisation.

The operation also highlighted some systemic shortcomings, which the French have undertaken to correct. Insufficient troop transport and tanker aircraft to project four companies and a headquarters is a problem that should be alleviated by an ambitious long-range transport procurement scheduled to support the Force d'action rapide. To be sure, the division will never have sufficient aircraft for rapid deployment of more than one or two regiments at a time, but that capability in itself will represent an enourmous improvement.

There are other *paras* in France. Those belonging to the 13th Dragoon Parachute Regiment have special operations missions in support of the 1st Army. The regiment (which, like those of the 11th Airborne Division and the army at large, are battalion-sized elements) has three companies, one with the corps stationed in Germany. The regiment has an exclusively European orientation and is principally tasked with long-range reconnaissance operations. It trains periodically with the German Fernspaehkompanien and, through that bilateral link, occasionally with other allied SOF. Unlike their German counterparts, though, for whom other operations such as sabotage or raids are not officially envisaged, the 13th Dragoons actively train in such commando tactics.

There is also the 11th "Shock" Parachute Regiment, formed in 1986 to carry out low-visibility protection, security assistance, and intervention operations overseas. The regiment is formed into corps of functional skills specialists, such as marksmen and demolitionists; groups of combat swimmers and divers; and a special forces element subdivided into commandos with both direct action and cadre-type missions similar to U.S. Special Forces. While the 11th "Shock" is an army unit, it is believed to be entirely controlled by France's intelligence service.

Counterterrorism

As in West Germany, the major impetus for creation of a national counterterror force in France came from the events at the 1972 Olympics in Munich; however, the relative inability of the French to deal effectively with the takeover of the Saudi Arabian embassy in Paris in September 1973 provided the final push.

The result was, and is, the Groupe d'intervention de la gendarmerie nationale (GIGN). The Gendarmerie is a military internal security arm, belonging to the Ministry of Defense. Thus, both army and GIGN utilization and procurement issues can be centralized. In practice, the GIGN operates as autonomously as most other national counterterror forces, whether police or military. Moreover, while the GIGN is technically not a police arm, it does have arrest powers and carries out operations against ordinary criminals as well as hostage takers. The unit exploits its military status for support, principally in overseas operations, but operates with the police.

GIGN is around seventy-five personnel strong, organized into five twelve-man teams and a headquarters. All personnel and their families live in quarters at a Gendarmerie compound near Paris, a measure that facilitates family member security and group member availability. Each twelve-man team has a team leader, two five-man sections, and a dog handler. GIGN depends on Gendarmerie or other military resources for most transportation, communications, and situation-dependent specialist personnel augmentation. However, it does have its own helicopters and a midsized executive jet to ensure its instant deployability. One team is on standby alert at any given time.

The GIGN's preferred mode of operations in a hostage incident is to eliminate as many adversaries as possible in a simultaneous sniper volley, and it will wait for hours on end for the one or two seconds when the optimum number of clear shots can be obtained. Should close assault become necessary, the single-aimed-shot principle still applies, with .44 or .357 Magnum revolvers being the main weapons. Automatic weapons are not normally used in such situations; however, when no hostages are involved, more firepower is employed.

A major instance in which this principle was applied was in Djibouti in February 1978 when thirty French schoolchildren were held hostage on a school bus by Somali terrorists. Ten GIGN personnel deployed to the scene, all taking up sniper positions. The commander then had mild sleep-inducing agents mixed in with the food being brought to the children to get them out of the line of fire in a manner that would not arouse terrorists' suspicions. His sniper volley, which was possible only after eight hours of waiting, killed four terrorists outright, and a fifth and sixth were killed as they exposed themselves. A seventh terrorist turned up on the back of the bus, and local forces engaged him in a firefight that killed not only the terrorists but one of the children. Hence the GIGN's dogma regarding selective, nonautomatic shooting.

Another feature found more strongly in GIGN than in other Western counterterror forces is its use of dogs. Each twelve-man element has its dedicated, trained dog handler. The GIGN has recognized that dogs can be a tremendous force multiplier in an apprehension effort, as well as providing detection capabilities and a psychological edge.

The GIGN has carried out over 250 other operations, securing the liberation or peaceable release of over 450 hostages. Its deputy commander and two other personnel were invited by the Saudis to take charge of operations to clear the extremists who had seized the Grand Mosque in November 1979 after previous poorly coordinated assaults had failed. The French team selected 90 Saudi troops (from over 3,000 available) and trained them intensively with gas masks and the use of gas dispensers. The, employing a nerve agent, the GIGN men led the Saudis into the mosque, where heavy subterranean vault-clearing combat ensued.

The GIGN is deployed overseas when French facilities or citizens are in danger. An assault team was prepared to reenter the French embassy in El Salvador in May 1979 had the situation not been resolved diplomatically. The GIGN intervened in New Caledonia when factional violence there resulted in the taking of police hostages in early 1988. Like most other major counterterrorist forces, the GIGN provides training for other nations. Except for the training agreement for the Saudis, which grew out of the Grand Mosque incident, GIGN assistance efforts have been on behalf of former French colonies.

The GIGN trains rigorously in physical and combat skills. Rather than create specialist teams, all personnel are required to be all things: parachutists, scuba divers, mountaineers, high-speed drivers, and, above all, marksmen. The aimed-shot gospel is reflected in the thousands of rifle and pistol rounds each man fires per year. Proficiency is also required in hand-to-hand combat, rappeling, and other military special operations skills. GIGN employs one unique training feature: each man is shot with a .357 Magnum while wearing a protective vest, both to accustom him to the shock he might expect to receive in an opposed assault and to instill confidence in his survivability.

The Future

The GIGN enjoys the full support of the government and is careful to foster a reputation for using firepower only as a last resort and only in circumstances with minimal risk to hostages. It has very little personnel turnover, and its equipment requests are, because of its small size and its authority to commandeer other military equipment when needed, comparatively modest. The future for the GIGN is probably secure, despite allegations in 1987 that its founder was involved in planting evidence on suspected IRA members and occasional misinformed attempts (generally by journalists) to link GIGN with those responsible for sinking the *Rainbow Warrior* in New Zealand.

Les paras, too, will undoubtedly survive, although there are some potential challenges to its force structure on the horizon. President François Mitterand has established his lack of enthusiasm for remaining "Africa's gendarme" and deployed forces to Chad only after much internal (and some external, mainly from the United States) pressure. He was so eager to terminate that deployment that he prematurely began to pull French forces out, suffering embarrassment when the United States demonstrated that Libya had not kept its part of the disengagement agreement. The government does recognize the requirement to be able to intervene if required, however—hence the Force d'action rapide (FAR), which is composed primarily of existing units, but some new manpower will be required, particularly in the airmobile division (in addition to airborne and airmobile divisions, the FAR consists of a marine, an alpine, and two light armored divisions). At the same time, however, the army is undergoing a 22,000-man reduction in force.

Thus, as the *paras'* overseas intervention mission is seen (perhaps wishfully) as increasingly less likely to occur, and one that will be shared with other forces when it does, the temptation may be overwhelming for the army to reorganize away some of the units' duty and strength. The same "less is more" fascination that currently has the U.S. Army staff reorganizing some support functions out of the 82nd Airborne Division (which is hard pressed now to support and sustain one brigade in a long-range deployment) and creating more but weaker divisions has already surfaced in French Defense Ministry visions of the future.

Reorganization or not and African disengagement or not, however, the skills and experience the airborne forces have acquired in over thirty-five years of operations in the Third World are critical to France's ability to influence events outside Europe. And in a European war, *les paras* comprise the most readily available fire brigade France has. It is not likely that France's airborne jacks-of-all-trades will be seriously affected by the upheavals attending the rest of the army's restructuring for the future.

Federal Republic of Germany

West German special operations forces and doctrines reflect its essentially defensive interpretation of its NATO role and a determination to keep the military responsible solely for protection of the Republic from external aggression. Its principal military SOF elements, the Fernspaehkompanien, are oriented toward missions involving stay-behind operations against forces that have passed beyond them or insertions behind enemy lines in enemy-controlled territory (versus the enemy homeland). Prehostilities infiltration into East Germany, for example, is (officially at least) out of the question.

Responsibility in the area of countering terrorism and other threats to internal order (such as the increasing menace of drug-related violence) rests with the Ministry of the Interior's Department P (Police Affairs). This department oversees Germany's paramilitary national police arm, the Federal Border Police, or Bundesgrenzschutz (BSG). Within the BSG is the nation's principal counterterror strike-protection force Grenzschutzgruppe-9 (GSG-9). Counterterrorism, at home or abroad, is viewed as a police, not a military, function, though the training GSG-9 personnel receive is heavily oriented toward military special operations skills, as well as generic close combat tactics.

Long-Range Reconnaissance

A long-range reconnaissance company (Fernspaehkompanie) is assigned to each German army corps. These companies perform reconnaissance missions principally, though their training enables them to perform sabotage missions if required. They are neither organized nor equipped for protracted offensive raiding or guerrilla support operations. At first glance, they may appear to have highly limited special operations utility; however, by specializing primarily in deep reconnaissance, they have brought such operations virtually to the level of an art form, and their operational concept much more closely parallels the strategic reconnaissance missions assigned to U.S. SOF than that of the mobile concept applied to U.S. long-range reconnaissance patrol (LRRP) companies in Vietnam. The Fernspaehkompanien are not expected to dash from one point on the battlefield to another, inserting for a few days and then being extracted.

They do not expect to have the luxury of absolute air supremacy (a requirement for such tactics) or the numbers of helicopters that a more mobile approach would dictate. More important, they believe interlaced, static hide positions provide a more consistent, reliable battlefield picture over the long term than do short-notice, short-duration point intelligence missions.

As a result, Fernspaehkompanien teams working for U.S. corps G-2 officers on NATO exercises at one time found themselves (as have U.S. Special Forces teams) misused by personnel who had no experience of directly tasking ground reconnaissance assets and who tended to think of them as Vietnam-era LRRPs. Because U.S. corps until recently had not long-range reconnaissance assets of their own (despite several attempts to have Special Forces given that mission), they also had no doctrine; thus ad hoc approaches born of misconceptions about capabilities (particularly in the area of mobility) and inexperience led to a tendency in many U.S. military intelligence quarters to downplay the importance of human ground reconnaissance efforts.

They would have, and should have, known better. In 1971, General Creighton Abrams, then the U.S. commander in Vietnam and no lover of Special Forces, nonetheless credited Special Forces cross-border reconnaissance teams with over 90 percent of his usable intelligence on North Vietnamese Army movements and capabilities. And so it has been for the German corps commanders who own the Fernspaehkompanien. During NATO maneuvers, they have had access to most of the same technical collection results as their U.S. counterparts, but they have usually credited their long-range reconnaissance teams with the vast majority of their combat intelligence.

The United States has, to some degree, come around to the German view on the utility of extensive stay-behind (or inserted) static reporting points. This is due partly to the previous NATO commander, General Bernard Rogers, and his view that technical collection platforms that survive a general Soviet onslaught may not be sufficient to provide the volume of continuous, reliable information on enemy movements that will be required. During his tenure, he successfully pressured a number of Special Forces commanders and trainers to concentrate on European mission orientations and techniques similar to those employed by the Fernspaehkompanien.

A second convincing element for U.S. corps commanders and intelligence officers was the performance of the Fernspaehkompanien and other allied elements in a joint long-range reconnaissance operation set up for the September 1982 iteration of "Reforger" (annual NATO war games involving deployment of forces from the United States). This operation, set up by a far-thinking U.S. corps G-2 intelligence operations officer, was scarcely mentioned in the mission planning sessions and briefings conducted back at the "Blue" Army Group headquarters. The conventional view considered the Joint LRRP Operations Center as a sideshow to the more serious business of tasking, collating data from, and analyzing the results of the technical platform missions (aerial photography, radar, and other forms of surveillance means). However, when the Red forces attacked, the technical platforms failed, for a variety of reasons: missions out of the United Kingdom were canceled because of adverse weather conditions, other missions were diverted to another command for political purposes, and so on. All of these represented possible actual eventualities that demonstrate the peril of overrelying on technical surveillance operations. It was the deployed long-range reconnaissance network that furnished the constant flow of bits and pieces that allowed the corps G-2 staff to detect a wide "Red" flanking movement. Initially these human-derived indications were not acted upon because of "insufficient technical source corroboration." Disaster for the "Blue" forces was narrowly averted by the late recognition that the human-derived information was, in fact, all too grimly accurate.

An upshot of this operation, and others like it, was that U.S. corps-level long-range patrol capabilities, once a throwaway mission given to a couple of National Guard companies in the United States (which would not be in Europe for thirty days after commencement of hostilities), are being revitalized by the activation of long-range reconnaissance companies as in-place corps assets. The example of the Fernspaehkompanien went a long way in bringing about that U.S. action.

The Fernspaehkompanien number around 140 to 150 personnel, or about 75 percent of their probable wartime strength. Each company is organized into a headquarters, two reconnaissance platoons, a communications detachment, and a support element. The reconnaissance platoons are comprised of a three-man platoon headquarters and ten four-man teams. Each team carries 9 mm silenced pistols and probably the HK-35 short assault rifle or MP-5 submachine gun (the standard German G3 assault rifle is considered inappropriate for the nonoffensive nature of Fernspaehkompanie missions). Each team is also equipped for FM voice communications and code-word burst transmissions.

Basic training for reconnaissance team members and company officers is conducted in the parachute and ranger courses at the German Army's Airborne School. Further training is conducted at the Long Range Reconnaissance Training Center, and unit training is often conducted with U.S., British, and Italian special operations units.

Counterterrorism

Although long desired by German security officials prior to September 1972, the concept of dedicated counterterrorist unit found full official support only when Palestinian terrorists carried out their massacre of Israeli athletes. The bloody failure of the Munich police to rescue the nine hostages taken by the terrorists prompted the decision later that month by the Ministry of the Interior to create a special unit within the Border Police. This unit was designated GSG-9.

GSG-9 is organized into four combat units of thirty personnel each (in the process of increasing to a strength of forty-two); a headquarters unit; a communications and documentation unit; an engineer unit; a training unit of thirty men (which can be utilized as another combat unit); a helicopter fleet of three helicopters and eleven pilots and mechanics; and a supply unit. The combat units consist of a five-man headquarters and five five-man special action teams. These teams are trained in protection of dignitaries, assault and assault support (sniping, grenade barrage), as well as such techniques as parachuting (including free-fall), underwater operations, high-speed pursuit and protective driving, rappeling, and hand-to-hand combat. This is a regime that exceeds that of the so-called SOF in some national military establishments. In addition, these teams receive a heavy diet of federal police and legal procedure subjects.

Their engineer teams are trained in underwater demolitions placement or removal, as well as the more conventional dry-land bomb disposal techniques.

Weapons for GSG-9 include the MP5 submachine gun (standard, silenced, and short-barreled versions), the G3SG/1 and Mauser 66 sniper rifles, and a variety of handguns. Rifle-mounted grenade launchers are available, and the unit utilizes an assortment of flash, stun, fragmentation, smoke, and gas grenades.

Organized and trained principally as a counterterror assault and rescue force, GSG-9 has been given other missions appropriate to a multicapable unit in a high state of readiness. Those missions include protection of designated German and foreign dignitaries, security assistance at high-risk German embassies, and augmentation of local and federal police forces in confrontations with increasingly better-organized and better-armed drug-smuggling and robbery gangs. This broadening of the mission base occurred both because the needs were valid and because the GSG-9 leadership has correctly surmised that bringing a force of young men through an exhausting training process to a high pitch of readiness only to do more of the same training can be counterproductive. The additional missions provide an opportunity to utilize skills and an outlet for energy.

The "father" and first commander of GSG-9, General Ulrich Wegener, is noted for putting into practice his belief that because terrorism is international, the war against it must be international. GSG-9 members have participated in the training of counterterror and protection forces around the world, notably those of Saudi Arabia, Somalia, Singapore, the Netherlands, Spain, and Austria. They have also participated as specialists or observers in most of the major terrorist incident resolution operations over the past decade, including the 1976 Israeli rescue at Entebbe and the 1978 SAS assault on the Iranian embassy in London.

The event that drew world attention to GSG-9 was its rescue

of eighty-seven hostages aboard a Lufthansa airliner seized by Popular Front for the Liberation of Palestine (PFLP) terrorists in October 1977 and landed at Mogadishu, Somalia. The occasional tendency of some laymen to compare this operation with the Entebbe raid is unfair to both the Israelis and GSG-9. GSG-9 was at Mogadishu operating in a permissive environment. The government of Somalia was more than willing to give the German government responsibility for resolving the incident and provided assistance to the GSG-9 effort in every way it could. The Israelis carried out their operation unilaterally in a hostile environment with only what they brought for support. However, the demands on GSG-9 for the measure of entry speed and surgical skill required in retaking an airplane full of hostages were no less intense because the outside environment happened to be friendly.

Rescue operations involving aircraft require intimate knowledge of the plane itself, split-second timing, a diversionary effort to concentrate adversary attention and make him move away from the entry point, and accurate, high-power "first-round, first-hit" shooting. GSG-9 achieved all but the last in its attack. With Somali troops setting fires in front of the cockpit and supporting commandos hurling flash and stun grenades (supplied by an accompanying British SAS team), the assault group moved up lightweight ladders placed against the tail section of the aircraft. On signal, they burst through the emergency exists.

The first man in engaged the nearest terrorist with a .38 caliber pistol, which did not have sufficient stopping power to prevent the terrorist from hurling a grenade. The grenade malfunctioned. The resulting explosion only wounded some hostages rather than creating the carnage it could have. The terrorist was hit with greater firepower a half-second later, as were his accomplices, but the lesson was a sobering one. In the end, three of the four terrorists were killed, the fourth was wounded, all hostages were recovered alive (though nine were wounded), and one GSG-9 trooper was wounded. The entire operation, from entry to complete evacuation of the plane, took seven minutes.

The Mogadishu operation had a significant impact on U.S. counterterrorism planning just as the German long-range reconnaissance doctrine would ultimately have in the corresponding area. While recruiting and training for establishment of the U.S.

dedicated counterterror force were already underway in October 1977, the performance of GSG-9 gave a new impetus to perceptions of urgency in both the Pentagon and the White House.

Developing program support among conventionally minded masters is never easy; however, the job does become a bit easier when one can relate a developing program to one developed by someone else to meet a similar challenge and point to concrete examples of success. In the areas of long-range reconnaissance and counterterror operations, the Germans have provided the community with two of the finest examples there are.

The Low Countries: Belgium and the Netherlands

Belgium and the Netherlands have long shared a perceived need to maintain combat-ready forces capable of conducting short-notice specialized missions. The origin of each nation's special operations capabilities can be traced to security considerations involving their former colonial empires and, more recently, the common threat of international terrorism. Those empires are now history, but the two North Sea neighbors have one further point in common; for each of them, the legacies of their colonial days have occasioned modern situations that required use of their SOF.

Beyond these very general similarities, however, there are few other commonalities in the two nations' appreciation of their overall special operations requirements or in the resulting force structure. The differences are rooted in divergent historical experience and national tradition.

Belgium's army fought in both world wars and developed a paratrooper force of exile volunteers in the second. Belgium never developed a maritime tradition, and its overseas colonial security requirements did not help foster one; the Congo had only one port and a narrow coast but a huge interior. Thus, in terms of continental military threat, colonial security requirements, and national role perception, Belgium developed what military tradition it has as a land power.

By contrast, the Netherlands has long had a naval tradition, and its farflung colonies in the East Indies required a naval and marine infantry presence more than one characterized by a large land army. Nor did European warfare experience alter that tradition: the Netherlands did not participate in World War I and was quickly overrun in World War II. The Dutch did fight the Japanese in the Far East, however, with (primarily) naval and marine forces.

Moreover, the Dutch lost their colonies through riots, insurrection, and outright seizure, and thus had no reason to feel any further responsibility toward them. The Belgians, however, retained extensive commercial and political interests in the Congo when they granted it independence. Thus, while the Netherlands no longer needed an extensive force projection capability, Belgium (like France) had to maintain the ability to intervene in its former dependency when required for the protection of its citizens and politico-economic position.

It is therefore unsurprising that the major military force element in Dutch special operations currently comes from its marine corps, while in Belgium it is a much larger army element. The Belgian forces' missions also cover most of the special operations spectrum of possible tasks, while the Dutch concentrate principally on counterterrorism and deep land and amphibious reconnaissance.

Belgium

Belgium's principal special operations establishment is its Paracommando Regiment, an airborne jack-of-all-trades force comprised of three battalions and supporting artillery, light armored reconnaissance, antitank, and administrative units. As such, it can (and does) operate much like any other conventional paratroop force; however, its commando training and orientation focus sharply on battalion-sized strike, interdiction, and point seizure operations similar to those envisaged for U.S. Ranger battalions.

The regiment is armed with a standard light infantry mix of rifles, pistols, machine guns, mortars, and antitank missiles. It also has a heavy infusion of submachine guns for close assault purposes. It operates a commando training center and a parachute training center in which it trains not only its own personnel but other elements in Belgium's special operations inventory, as well as small contingents from other allied special forces units.

The Paracommando Regiment has had two opportunities in its

postwar life to perform its mission, both in Zaire. The first was in 1965, when Simba rebels fighting the Western-supported central government had overrun several towns in the northern part of the country. Several hundred European civilians were captured, and it required no gifted imagination to picture what would happen to them; recent history in the political-tribal upheavals both in Zaire and elsewhere had provided ample grisly examples of the Europeans' probable fate.

The troops of the Paracommando Regiment were hastily assembled and provisioned, given scanty instructions, flown south, and dropped to help head off the anticipated massacres. They linked up on the ground with the central government's hard-driving mercenary columns and with them raced into the beleaguered towns. The speed and violence of the combined attacks were such that the Simbas were surprised virtually at the outset of their intended mass raping and slaughter; hundreds of Simbas died, and the vast majority of imperiled civilians were recovered, badly shaken but unhurt.

The second deployment was also a rescue operation, mounted in loose cooperation with the French in May 1978, during an incursion by Shaba rebels into their former homeland. It was originally envisioned in the hasty, inadequate joint planning session that launched the operation that the French and Belgians would secure key areas within respective halves of a large perimeter, near the overrun city of Kolwezi, and then go on to liberate European civilians being held by the invaders whereever they were found. However, the earlier arrival and violent assault operations of the French Foreign Legion paratroops had reduced remaining resistance in Kolwezi to isolated pockets by the time Belgians arrived. The French after-action report for the operation mentions difficulties in on-the-ground coordination but does not elaborate; other sources indicate sharp differences between the commanders on the speed and aggressiveness the Legion was using in its city-clearing operation and in methods used in interrogating prisoners. The upshot was that the French turned the task of mopping up in and around the city, securing the airfield, and processing rescued civilians over to the Belgians, while they went off into the bush to hunt down remaining rebel units.

The Belgians accomplished their part of the operation with competence and professionalism, if not with the élan of the Foreign

Legion. The difference in approach to the mission is not a difficult one to assess, and the Belgian approach does the paracommandos no discredit. Both forces came to rescue people, but for the Legion it was first and foremost a combat operation. During the rescue phase, speed counted, and the French approach probably saved many lives. Had the Belgians been in the same position, it is probable that they would have reached similar conclusions. But for the Belgians, the killing of Shabans served a purpose only insofar as finding and protecting the Europeans required it. For the Legion, the Shabans were an enemy force, upon which as many casualties as possible should be inflicted before the rebels succeeded in getting back across the Angolan border.

Beyond such considerations of operational style, however, both of the Paracommando Regimant's outings demonstrated two things: that the regiment could quickly assemble and outload a combat-ready battalion-sized force for long-range deployment on short notice and that neither deployment could have occurred without the U.S. Air Force. Both conditions remain true today.

In addition to the Paracommando Regiment, the Belgian Army has a special reconnaissance company assigned to its NATO-committed I Corps. This company is tasked with NATO-particular traditional unconventional warfare missions like behind-the-lines sabotage, aid to resistance elements, and strategic reconnaissance. It is organized into a headquarters element, two operations elements, and a multisize-capable signals unit. It undergoes training at the Paracommando Regiment's parachute and commando schools, conducts language and some demolitions training on its own, and does high-altitude, low-opening parachute training with other allied units.

Belgium's national-level counterterror force is the Escadron special d'intervention ("special intervention squadron") of the Gendarmerie. Like its French counterpart (with whom it has done some training), its Gendarmerie status puts it under the Ministry of Defense from whence it draws various types of support and gives it police powers as well. Also like its French counterpart, it performs some dignitary and embassy protection functions. It has carried out other more routine police-type functions. At one time retention and morale were low, principally because of low pay, no extra benefits, long hours, and a surfeit of missions. The Escadron has not yet had the kind of major incident to contend with that helps forge traditions, gives focus to training, and builds a broad-based official constituency. However, the outbreak of anti-NATO terrorist violence that began in 1984–1985 and will undoubtedly increase if Belgium deploys U.S. cruise missiles has raised the Escadron's official visibility as well as its activity level, and it is now receiving a more appropriate level of dedicated funding support.

The Netherlands

Like the Belgians, the Dutch have a reconnaissance company assigned to their national corps NATO contingent. This long-range patrol company is the remnant of several former commando companies, now disbanded. Although members still receive commando training and the unit can therefore (like its West German counterparts, which it resembles much more than it does the Belgian company) be presumed to have some sabotage and raid capabilities, such operations are not acknowledged as comprising part of the mission profile. As with the Germans, this company specializes in deep, detailed reconnaissance operations. Its teams train to operate from "hide" positions far from friendly lines and to be self-sustaining. It is organized into a headquarters, a signal platoon, and four operational platoons and is regarded highly.

Where the Belgians' principal contribution to NATO beyond its mechanized conventional land force is the Paracommando Regiment (one battalion is on call to the Allied Command Europe Mobile Force), the Netherlands' is, appropriately, oriented toward the sea. The bulk of its active Royal Netherlands Marine Corps is organized into the 1st Amphibious Combat Group, structured, equipped, and trained in a manner consciously designed to be as identical as possible to a British Royal Marine commando. This facilitates its wartime NATO mission of operating as an integral part of a British Royal Marine brigade group (3rd Commando Brigade) on the Northern Flank. To support that mission, marine reconnaissance personnel carry out amphibious and underwater operations similar to those performed by the SBS and SEALS.

The major Dutch special operations unit is a company of the marine corps known as the Special Assistance Unit—Marines (or, using the Dutche initials, BBE). It is trained for close-combat assault operations and could be used against military targets in wartime. Indeed, the company is an integral part of the 1st Amphibious Combat Group and deploys with it on major NATO training exercises. However, its primary mission in peacetime is counterterrorist operations, and at least one platoon is always on standby in the Netherlands.

The BBE has three operational thirty-man platoons, organized into five-man teams. Usually only two platoons are on duty status, while the third is undergoing a year's intensive training. Any new recruits are trained by this platoon; then the platoon as a whole undergoes refresher and advanced training in conventional military skills, commando operations, close combat, and other counterterrorism-specific subjects. The BBE can call on a backup of sorts in the form of an organized-as-needed reserve platoon composed of former unit members serving elsewhere in the Dutch forces. The BBE is armed with HK submachine guns and automatic pistols and has the usual inventory of stun, gas, smoke, and explosive munitions.

The BBE has had several opportunities to display its proficiency. In October 1974, four convicts took over a prison chapel in The Hague, holding nearly two dozen hostages. In the midst of negotiations, a BBE platoon stormed the chapel and surprised the convicts. All the convicts were arrested, the hostages released, and no one was harmed.

A more complex situation faced the BBE in May 1977 when South Moluccan terrorists simultaneously seized a train and an elementary school in a nearby town. Negotiations continued for over two weeks. Finally the BBE was called upon to make the two assaults, backed up by other police and military elements. The assault on the school was accomplished in two raid stages.

Several armored vehicles formed up in front of the school as a diversion and to provide cover for the BBE's final deployment. Four other armored personnel carriers smashed through a side glass wall, through which the following BBE troopers fired gas and then charged. All the terrorists were captured. In assaulting the train, another BBE platoon had the advantage of knowing exactly where the terrorists were located because of listening devices and other surveillance gear surreptitiously placed under and around the train during the prolonged negotiations period. In the predawn darkness, the platoon deployed as close as possible to the train, backed up by dozens of police and military snipers. Then, simultaneously, Dutch Air Force F-104s carried out ear-shattering low-pass simulated bomb runs, explosives simulating the bombs were set off, and snipers took various points on the train under fire. With this din creating a diversion, the marines blew out preselected window areas, entered the train, and killed or captured all the terrorists with only slight friendly casualties.

In March the following year, more South Moluccans held seventy hostages in a provincial government complex. Once again the BBE made extensive use of painstaking surveillance and route analysis. This time no diversions were used; rather, the BBE platoon entered from adjoining buildings and moved unobserved to an area outside the room in which the hostages were kept. With a single grenade blast to knock down the locked door, the BBE moved in, capturing the three terrorists and rescuing the hostages.

Conclusions

The differing character of emphasis placed by each of these geographic and political partners on their national special operations capabilities is a function of their different histories and the challenges they have faced to date. Both nations appear comfortable with their particular emphasis and have ensured that it serves both national and NATO requirements. The Belgian Paracommando Regiment and the Dutch BBE have correspondingly different mission profiles and training priorities. But they have one more point in common: more than once, both have validated their nation's faith and pride in their capabilities. No SOF can aspire to any achievement more meaningful.

The Northern Flank: Norway and Denmark

The special operations establishments of NATO's Northern Flank countries are, like the military force structures from which they are drawn, a good deal smaller than those of most other NATO allies. Even small countries like Belgium and Holland, with current political climates not unlike those of their Scandinavian brethren, maintain more extensive special operations capabilities in spite of being somewhat farther removed physically from the Soviet threat. What the Scandinavians lack in numbers, however, they have apparently determined to try to make up in specialized expertise.

Nowhere else in NATO is deep reconnaissance in winter conditions performed better than among the Norwegian Parachute Jaegers, and the Royal Danish Navy's Fromandskorpset (Frogman Corps) is renowned throughout the NATO special operations community as one of the best underwater demolitions and amphibious reconnaissance organizations. A closer look at the capabilities and operational philosophies manifested in the two countries' approach to special operations reveals a number of shortcomings (primarily budgetary) but also how closely those approaches reflect national tradition, historical experience, and current perceptions of how best to contribute to NATO and national defense.

Denmark

Special operations troopers from the United States, Great Britain, West Germany, and the Low Countries are familiar with the Danish "scout swimmer" (amphibious reconnaissance) school, the best, many believe, of its genre in Europe. The school and its training reflect the dedication and professionalism of the Fromandskorpset. They also reflect some of its limitations; when the Fromandskorpset is running a training course of its own, it lacks manpower to do much else. Numbering fewer than fifty personnel, it draws volunteers from all over the Danish armed forces. The successful candidates are transferred to the navy if not already in that service.

The Fromandskorpset trains for a number of missions. Given Denmark's geographical position astride the approaches from and to the Baltic Sea, emphasis on mine planting and mine clearance, underwater demolitions, and rescue and recovery is not surprising. Other diving tasks, such as salvage, obstacle clearance, and amphibious reconnaissance, are also assigned. Sea-launched offensive operations on the ground have been, until recently, a secondary mission. However, sharing concern with other North Sea countries over the threat by terrorism to oil rigs and servicing vessels, Denmark has tasked the Fromandskorpset with maritime counterterror duties as well. Only time will tell whether adoption of this counterterror mission can be accomplished without spreading assets so thin that overall capabilities are degraded. It is known that the Fromandskorpset performs its other missions with high proficiency.

The Fromandskorpset aggressively seeks out joint training opportunities with other NATO SOF, not only in the cause (laudable enough in itself) of interallied operability but also in the more pragmatic pursuit of badly needed training that the Danish government cannot fund. By offering joint training to the United Kingdom or the United States, for example, the Danes hope those countries will participate with their considerably more diverse seaborne or airborne launch platforms and support equipment. Everyone gains, and the allies cheerfully submit to being used in this manner by troops whose skill and spirit they admire. An example of training that can seldom be conducted without U.S. or U.K. aircraft support is a parachute operation launching both personnel and a combat rubber raiding craft.

Less well known outside Europe is the Danish Army's Jaegerkorps, or special reconnaissance company, which has a headquarters and four ten-man patrol units. Like the Fromandskorpset, it suffers from an overabundance of diverse tasks and a paucity of personnel and equipment resources with which to perform them. However, like their maritime counterparts, the Jaegers have demonstrated great enterprise in obtaining training through the good offices of their allies that would be difficult or impossible to achieve unilaterally. At one time, the Germans conducted parachute training for the Jaegers, and currently the Jaegers assiduously absorb techniques from British and U.S. special operations participants in annual NATO-wide unconventional warfare exercises. Some skills are best learned at home, however. Denmark's geography and the hard probability that in any general conflict much or all of the Jutland Peninsula may be overrun, have dictated some of the Jaegers' training priorities. The first is that, despite their army support mission, the Jaegers are likely to have to move clandestinely by amphibious or underwater means. Thus, they are trained in amphibious and underwater skills by their counterparts in the Fromandskorpset (they return the favor to the frogmen with parachute training). The second training priority involves training for and with elements of the Danish Home Guard. In wartime, the Jaegers and the Home Guard would be mutually dependent on each other in operations behind enemy lines. U.S. and U.K. SOF also train with the Home Guard.

The Jaegers have also been given a counterterror mission. As with the Fromandskorpset, resources and training time that can be devoted to this mission are limited and may be insufficient to the task unless other missions are degraded. Allied observers uniformly agree that Jaeger performance in those other mission areas is highly professional. This proficiency is laid to the aggressiveness in pursuing skill-enhancing training wherever it is offered and to the Jaegers' composition (all members are long-term career soldiers, volunteers who will spend most of their army careers in the unit).

Clearly Denmark believes its unconventional warfare priorities are in supporting its primary conventional goals: closing the Baltic approaches (with allied help) while maintaining internal over-water lines of communication and sustaining that closure through a combination of Danish Army (and allied) defensive operations and Home Guard stay-behind activities. To those priorities has been added a peacetime counterterrorism role.

Both forces must maintain skills for which the Danish government provides insufficient training funds and equipment resources. If assignment of the counterterror mission is not accompanied by personnel increases and some more and better equipment, the Fromandskorpset and the Jaegers may be hard put to maintain true proficiency in their overly broad mission spectra.

Norway

While the World War II resistance tradition in Denmark was necessarily of an urban-oriented sabotage and espionage variety, Norwegians operated with British commandos in both the fjords and in overland ski marches. Today Norway hopes to hold most of those fjords until British, U.S., and Dutch marines arrive to break the back of any Soviet threat. To help hold the fjords or deny their use and to facilitate Home Guard resistance in the Soviets' rear (primarily in the north) are the principal thrusts of Norway's special operations approach.

The Norwegian counterpart to the Danish Fromandskorpset is the marine Jaeger platoon. It is trained in underwater demolitions, use of collapsible boats, kayak operations in fjords, and amphibious reconnaissance techniques. At one time, it trained extensively in submarine-launched operations.

Like their Danish counterparts, the marine Jaegers train often with their maritime allies (United States, United Kingdom, Denmark, and the Netherlands) not only to further interoperability, a wartime necessity, but to avail themselves of equipment and operational scenarios not otherwise realizable in unilateral training. However, this problem is not as pronounced as it is in Denmark, for at least two reasons: the Norwegians manufacture a good deal of the equipment the marine Jaegers need for fjord and winter operations, and the unit is quite small (fewer than fifty personnel).

Nevertheless, the marine Jaeger unit reportedly spends most of its time training for its new role as the amphibious half of the military's new joint counterterrorism unit. While observers believe that such training is being professionally accomplished and achieves such general benefits as small unit tactical expertise and cohesion, it is being conducted at the expense of training in the more traditional maritime special operations skills.

The army component of Norway's small special operations establishment is the parachute Jaeger group. Its basic element is the five-man team, which operates as a self-contained element in unilateral patrol or strike missions or as part of a training cadre in behind-the-lines operations with in-place Home Guard resistance elements. The parachute Jaegers train regularly, albeit in small elements, with British and U.S. SOF and with the Norwegian Home Guard. Like the Danes, the Norwegians expect to support, and be supported by, stay-behind resistance elements during any general conflict, and the parachute Jaegers (and allied counterparts) are integral to this element of Norwegian strategy. This training reflects appreciation of the operational requirements of that strategy in some respects. Blessed with their air force's own cold-weather-capable C-130s, the parachute Jaegers can and do spend considerable time practicing high-altitude parachute insertion techniques. They also have honed cross-country ski patrol and arctic survival techniques to a fine edge. Other individual and small unit skills receive similar emphasis.

Budgetary and social factors, however, impair the Jaegers' ability to mount and sustain an integrated wartime special operations effort. Their actual duty strength is reliably estimated at only a third (about one hundred personnel) of their originally authorized manning level. Unionization of the Norwegian military, which has provided for competitive salaries for those in uniform, has also tended to force the government to limit the numbers of those individuals. Moreover, overnight or weekend training commands higher rates of pay. Thus field training of the type that exercises the Jaegers' internal operability and centralized control of a broad reconnaissance and unconventional warfare effort under extended field conditions is exceedingly rare.

Most recently, another constraint—the counterterrorism mission—has been added, which competes for already-scarce training time and resources. The parachute Jaegers provide elements that make up the army half of the joint counterterrorism unit. This unit was formed in the early 1980s to augment the existing domestic counterterror capabilities (then held solely by the "Readiness Troop" of the Norwegian National Police) and extend them to protection and response operations involving Norway's slice of the North Sea oil rigs.

In a counterterror situation, the military unit will be under the operational control of the police supervisor on the scene, who in turn takes orders directly from the Ministry of Justice and Police.

The Norwegian National Police Readiness Troop is organized into four twelve-man squads, each of which is then broken down into a small leadership element and five two-man teams. The military unit is reported to have a headquarters section (which includes bomb specialists and a medic) and six five-man commando squads. Neither unit has been tried in an actual crisis, but the realism of the training that does take place is reportedly high, and performance of the Readiness Troop is said to be very good.

Results of the military's counterterror training have been only scantily observed, but what reports there are indicate the troops are going at it with a will. The major problem, voiced by some leaders and evident to observers, has been addressed throughout this assessment: a set of missions and national expectations that are not met by commensurate manning levels or training budgets.

Conclusions

The governments of NATO's Northern Flank appear to have appreciated the roles and mission priorities for their special operations establishments that best contribute to both national and NATO defense goals. They have also appreciated, as have so many other countries, that the kind of dedication and small unit cohesion that characterizes most SOF serves as fertile ground for development of or strengthening of national counterterrorism capabilities. The dividend achieved thereby—peacetime utility for an otherwise expensive capability usable only in the war nobody wants—is not lost on the Scandinavian governments either.

What neither government apparently appreciates is that the resources they have allocated for their special operations establishments to perform their missions are significantly less than those required. While to some degree this is true of most other special operations establishments in the West (including those of the United States), the proximity of both nations to the source of their principal threat argues strongly for a reassessment of priorities. Both nations can take some comfort in their rapid-mobilization reserve systems. The Norwegians in particular have created an excellent system that enables the individual Home Guardsman to retain his weapon and equipment at home and thereby report in crisis immediately to his unit's battle position, without mobilization assemblies. However, that means of rapidly augmenting frontline and behind-the-line fighting strength cannot currently be quickly supported by the special operations establishments on anything like the scale required. Moreover, reserve special operations resources are slim in both countries. Addition of the counterterror mission in both countries—an eminently justifiable use of special operations units if adequately supported in all respects—exacerbates already critical training and manpower situations.

Both nations' defense planners have thought through the appropriate roles for their special operations units. What they now must think through—and convince their governments of—is the reality that the resources assigned to those units will not support performance of those roles.

The Southern Flank: Italy and Turkey

Special operations on the Southern Flank of the Atlantic Alliance, represented principally by the forces and doctrines of Italy and Turkey, are a study in contrast. Each nation's perceived requirements, and the capabilities developed to meet them, reflect unique national characteristics and geopolitical realities. The only similarities between the two nations' forces and doctrines are those that result from common usage of various weapons and techniques and from decisions recognizing both Alliance and national needs for a range of special operations skills and the forces to operate them. Both nations, for example, maintain forces trained to support guerrillas and partisans, both operate counterterror forces, and both can deploy direct action forces for raids and sabotage; however, the degrees of emphasis and the shape of the resulting force structures in the two principal Southern Flank powers are quite different.

Italy

Italy's primary unconventional warfare unit is the Ninth Airborne Assault (Saboteur) Battalion of the Folgore Airborne Brigade. It is one of four battalions in the brigade and conducts training at the command and staff levels in operating as an integral maneuver element. Its personnel participate with other brigade members in conventional-operations-oriented exchange training visits with the U.S. 82nd Airborne Division and other allied airborne forces. Its unit training, however, is oriented on individual and team skills much like those employed by U.S. Special Forces. Emphasis is placed on underwater operations, airborne-launched and overland sabotage operations, deep reconnaissance, and support for partisan or stay-behind conventional forces. Its missions are primarily related to support of Italy's NATO role and have included mountain operations, although the army's mountain brigades also have limited reconnaissance and sabotage capabilities. Much of the training is essentially on the job and takes several years to complete. All battalion members are volunteers, and its enlisted personnel reach noncommissioned officer rank.

Proficiency in the individual skills of the special operations trooper is rated high by other allied special operations observers, for that is where the training emphasis is applied. They are considered particularly good at high-altitude, low-opening operations. Ability to conduct sustained unit operations, however, particularly in a partisan support environment, is rated less enthusiastically. These rather more complex activities integrating command and control across a broad spectrum of conventional and unconventional conditions are not exercised often.

The Italian navy's special operations contribution is the Comando Raggruppamento Subacquei ed Incursori, loosely translated as the "frogman and raiders' group" and known as COMSUBIN. The missions of COMSUBIN are similar to those of U.S. Navy SEALS and include mine and underwater obstacle clearing, amphibious reconnaissance, and surface and subsurface raids on enemy ships, port facilities, and fuel storage areas. Organizationally COMSUBIN is split into a raider operations group for offensive operations and a frogman group for Italian coastline support. In practice, personnel from both groups will comprise a multicapable task element.

Personnel are drawn primarily from Italy's small marine corps, the San Marco naval infantry battalion. That battalion itself receives general commando training from COMSUBIN, but when its personnel volunteer for COMSUBIN, they leave the San Marco battalion. Basic underwater instruction for the frogman group's personnel is conducted over most of a year. Those going to the raider group complete the program training and then proceed to a further year's training, which incorporates typical commando skills such as sabotage, parachuting, and small unit raid operations with rigorous physical endurance tests and mountain climbing.

COMSUBIN is reasonably well regarded among naval special operations observers. The Italians have an extensive coastline and are assuming an increased role in NATO Mediterranean naval operations with the launching of their first aircraft carrier. COMSUBIN's training and skills reflect Italy's appreciation that, while its land special operations efforts are likely to take place in the North and Northeast, its naval elements must be able to operate in a variety of climates and operational settings.

Both the 9th Airborne Assault (Saboteur) Battalion and COMSUBIN have been designated as counterterror forces by the Italian government, reflecting the government's decision to spread the counterterror charter around among military arms in addition to that part of the mission given to the Carabinieri. The paratroops are unquestionably imbued with the esprit and individual troop talents to take on the mission as a contingency and have sent several dozen unit members to train with Great Britain's SAS. However, as with the Nordic countries, which attempt to levy all conceivable special operations missions on one, small force structure, it is doubtful that the 9th Battalion can address all those missions with equal emphasis. Fortunately, Italy has not given Folgore the heavy burden of principal domestic counterterror responsibility.

Nor does COMSUBIN have that role, though unlike the 9th Battalion, it has been utilized in actual counterterror deployments at home and abroad. It was called out when a hijacked airliner from Beirut was brought to Rome in September 1979, and it has reportedly been involved in either augmentation or on-the-job training with undercover police operations. Most recently, while much of the media coverage during the *Achille Lauro* affair speculated on U.S. force locations and plans, COMSUBIN elements were deployed on Italian vessels near Lebanon, indicating government acceptance of at least the possibility of COMSUBIN personnel being used to resolve terrorist incidents in international waters.

At the same time, however, the Gruppi Interventi Speciali (GIS) ("special intervention groups") of the Carabinieri were also reportedly represented on the floating reaction element, indicating that in action involving internal Italian jurisprudence (which included the *Achille Lauro,* an Italian flag vessel), the Carabinieri constitute the lead agency. Certainly in the major domestic counterterror campaigns—the nationwide fishnet for the kidnappers of former Prime Minister Aldo Moro in 1978, the Trani prison takeover resolution of 1980, and the search for kidnapped U.S. Brigadier General James Dozier, a year later—the GIS have played prominent roles. At Trani, the GIS encountered a prison entirely controlled by inmate terrorists and other criminals who had been supported from the outside. The GIS rigged demolitions at the gates and at other points on the walls; landed troopers on the roof; blew the demolitions both as a diversion and to facilitate entry by a ground element; and then penetrated through the roof, rescuing eighteen hostages and subduing the rebellious inmates. It was not a bloodless or perfect operation; nearly two dozen commandos and a small number of hostages received some wounds, largely results of the various demolition blasts. But the assault gave the nation, which at that time had been virtually held in thrall by the Red Brigades, a major shot in the arm.

GIS personnel are recruited from all over the Carabinieri but particularly from its unique airborne battalion, which operates with the Folgore brigade and has the mission of working with the army in countering insurgency and organized crime in difficult terrain. GIS training reflects ties with Germany's GSG-9 and the SAS. Preferred operational techniques include short bursts from Beretta 93 machine pistols, backed by Beretta SMGs. The GIS also use Mauser Model SP 66 sniper rifles and a variety of optical and night vision equipment.

Italy has appeared to ride a pendulum on application of energetic measures to combat terrorism. The kidnapping of Aldo Moro triggered a nationwide manhunt, which, while unsuccessful, netted a great many Red Brigades rank and file and supporters. Imprisoned as ordinary criminals, no special visiting restrictions were applied to them, and outside supporters were able to provide the Trani inmates with a variety of weaponry. The takeover and the raid were the results. Activities by the Red Brigades dropped off, as did counterefforts by the Italians. When Brigadier General Dozier was kidnapped, the GIS "leatherheads" did indeed provide the denouement to weeks of painstaking investigation. But the real victory was the discovery and capture of a number of Brigatisti weapons stockpiles and action cells that occurred in the course of the search. And the real accompanying defeat was threefold. First, relaxed preventive investigation had allowed the Brigades to rebuild to unsuspected levels. Second, it took a high-visibility trigger like Dozier's kidnapping to bring about the productive crackdown (as it did in the case of Aldo Moro). Third, by all accounts that came in from the fallout of the search for Mohammed Abbas's accomplices, the Red Brigades have again been rebuilding.

Questions of accommodating the United States aside, it is interesting to note that the government's initial reluctance to hazard its rapprochement with the PLO over Abbas apparently ignored the fact that many of the arms caches found during the search for Brigadier General Dozier were traced to Palestinian sources. Thus, the "leatherheads" seem assured of steady employment in the future.

Turkey

The Turks are unique in the Atlantic Alliance in having dedicated their principal military special operations capabilities for the specific purpose of organizing, training, and supporting an in-place guerrilla infrastructure. With border regions facing the Soviets and

Bulgarians and military forces poorly equipped in comparison to potential Communist adversaries, the Turks have reckoned with the probability of at least the temporary hostile occupation of those borderlands and possibly even more territory. They have determined to make that occupation costly to invaders not only in nuisance and diversion terms but in organized attacks from the rear coordinated with operations at the front.

The Turkish army's Special Warfare Department (SWD) is set up in several elements, all designed to provide complete training, administration, logistical command and control, and conventional headquarters liaison functions for a number of predesignated unconventional warfare operational areas (UWOAs). The headquarters element provides centralized operational, intelligence, logistical, and communications support and direction and PSYOPS support. Training for both regular and active duty–reservist cadres is run by a small training group, which also generates the doctrine and materials used in ongoing field training in the UWOAs.

That training is conducted for the reserve units in the UWOAs by a small number of active duty cadres modeled on U.S. Special Forces A teams. Indeed, at one time, the U.S. Special Forces ran an almost continuous variety of MTTs with the Turks (as they did with the Greeks), but Turkish reaction to U.S. congressional expression of disapproval over Turkey's occupation of part of Cyprus greatly reduced that cooperation in recent years. The reserve guerrilla units have their own chiefs but are linked to the SWD by regional special warfare staffs in each UWOA. These staffs also serve as liaison elements with the local major conventional army unit.

What the Turkish training lacks in equipment sophistication and modernity, it makes up for in mountain operations–oriented ruggedness. The Turks consider that the grit and soul of its soldiery have always been its principal military assets, and while Turkish military leaders would doubtless wish to see greater firepower potential for all their forces, conventional or otherwise, they have clearly decided on a course that they feel makes the most of their strengths. Whether that qualitative factor can overcome the lowest level of military modernization in NATO remains to be seen.

The same approach—overcoming deficiencies in technique and equipment with an abundance of ferocity—typifies Turkey national counterterror effort. Turkey has a commando brigade with both offensive raiding and internal security responsibilities, but the principal counterterror force belongs to the Jandarma. The Turkish word for their designation means, in context, "self-sacrificing"; more literally, it means "suicide," and the unit's troopers are colloquially known as the suicide commandos.

The commandos receive individual training in most special forces specialties, as well as those more appropriate to a paramilitary police unit. Of particular interest is emphasis on heliborne and chemical operations. They train in unit close assault and traveling security operations and believe speed and violence of action are the keys to successful incident resolution. Emphasis on preserving the lives of innocents in an incident-reduction situation does not appear to be any less than that of similar European forces, although killing or capturing the terrorists is the principal mission.

The unit has had two major outings in the recent past. In one, it resolved an aircraft hijacking by employing noise and light at the front of the plane as a diversion and then staggering break-ins through doors at the tail, in front, and through a small door behind the cockpit. The terrorists were thus continually and successively engaged from different directions and could not coordinate their response. The second involved Armenian seizure of an airport, which the commandos resolved when the terrorists made good on a threat to begin blowing the place up.

The first operation was a complete tactical success; the second saw the annihilation of the terrorists, to be sure, but the demise of many civilians as well. Whether either operation could have been improved with more sophisticated equipment and training is difficult to say. The point can justifiably be made, however, that a nation that consciously sends high-quality human material into highly dangerous situations for the national good does them (and any civilians whose lives may depend on their performance) a disservice if the equipment and training do not match the spirit. Much of the doctrine and all of the equipment of both the SWD and the suicide commandos are antique by contemporary standards. Clearly a nation that takes as much pride in pulling itself into the twentieth century almost overnight after World War I as it does in its people's martial qualities can and should commit itself militarily to completing the former process in order to optimize, rather than waste, the latter asset.

3
The Role of Special Purpose Forces in the Soviet Union: The *Spetsnaz*

> The telephone rang. [The NATO Northern Army Group duty officer] answered it and heard the familiar voice of the British Liaison Officer at I German Corps. It was urgent and agitated. "Parachutists!" it said, "Russian parachutists on . . . " The voice abruptly died.
>
> Several direct-line telephones were jangling in the Ops room at once. He heard shouts in the building. There were shots outside, the noise of helicopter rotors, a violent explosion.
>
> He threw the switch of the alarm system and unhooked the direct line to AFCENT and in that moment looked up. A Russian soldier was standing in the doorway. The Duty Officer was getting to his feet and reaching for his pistol . . . when the Russian shot him dead."
>
> —General Sir John Hackett et al.,
> *The Third World War: August, 1985*

The potential reality of this scenario and others like it involving not only headquarters but communications, fuel, and transportation arteries, as well as ammunition storage facilities, airfields, and nuclear weapons storage sites, has developed into a major concern for Western military planners. Once dismissed as nuisance level, the qualitative and quantitative capabilities of the Soviet Union's collected "special purpose forces," known generically as *spetsnaz,* represent what is now seen as part of a serious Soviet intention to inflict major damage on their opponents' C^3I and operations-sustainment infrastructure. This threat to rear area security and forward operations lifelines has resulted in dramatically increased training for Western military police, airfield security, and other support-type units in protecting critical facilities, countering raids, and resisting convoy ambushes. Most of that emphasis has been reflected in doctrinal planning, mission taskings for NATO home guard units, and small exercises featuring attempted penetration of facilities by friendly SOF.

Great Britain went several steps further in 1985 when it conducted Exercise Brave Defender on a country-wide basis. The scenario involved sequential multisite attacks by aggressors representing the *spetsnaz* against the U.S. and British air bases, command and control centers, and other designated targets. In defense were primarily the mobilized forces of the TAVR and the Royal Air Force Regiment (tasked with defending air force fields), with some participation from U.S. Air Force security police units. Over 5,000 SAS, Royal Marines, and other troops played the attackers, and over 30,000 personnel took part in counterintelligence, search, and installation defense.

The United States, too has factored *spetsnaz* known capabilities and potential plans into its own domestic defense exercises, though on nothing like the scale and with nothing like the centralized impetus as Brave Defender. Exercise Brim Frost, biannual winter war games in Alaska, have more than once featured U.S. Army Special Forces and U.S. Navy SEALS in the aggressor role, with targets including not only military airfields, communications, and POL storage facilities but sections of the Alaska pipeline as well. The navy and the Coast Guard have conducted protection exercises in the Chesapeake Bay with army reserve and navy SOF playing the *spetsnaz* role. And in the largest multiagency internal security exercises ever conducted in the United States, the Department of Energy's Strategic Petroleum Reserve has on two occasions postulated *spetsnaz*-level prehostility threats and countered them with integrated intelligence and reinforcement operations involving numerous local, state, and federal law enforcement and military agencies. Other Department of Energy sites have conducted similar though much smaller exercises, and the Department of Energy's principal security official has instructed field offices to include *spetsnaz* capabilities in their overall facility protection planning.

The "*spetsnaz* awareness" factor and the response in Western defense planning circles are relatively recent phenomena. The threat that generated them, however, is not new at all. The KGB and the GRU (the Soviet military intelligence arm) have long had destabilization, sabotage, strategic reconnaissance, and assassination missions and capabilities. *Spetsnaz* elements infiltrated into Czechoslovakia just prior to the May 1968 invasion and facilitated the occupation of Prague airport by the 103rd Guards Airborne Division. A GRU *spetsnaz* element, acting under KGB direction, engineered the December 1979 coup and assassination in Afghanistan that brought Babrak Karmal to power. Four-man teams operated briefly in Vietnam in the late 1960s and early 1970s to field-test the new Dragunov sniper rifle.

Closer to home, creditable, albeit isolated, reports have had *spetsnaz* advisers working with Cuba's *Commando Navale,* with Nicaragua's six counterinsurgency battalions and its air base defense forces, and even in training demolitions specialists in El Salvador's Marxist guerrilla movement, the FMLN.

The GRU "owns" the Soviet military *spetsnaz* formations and exercises its control both centrally, as in the case of detachments and agents operated directly out of the 2d Chief Directorate of the General Staff, and through a *spetsnaz* group at the front and military district-level staff. In the latter case, the front *spetsnaz* group operates under the staff 2d Directorate. It controls a detachment of *spetsnaz* intelligence operatives, whose mission is to gather and maintain intelligence on *spetsnaz* strike targets, and a *spetsnaz* brigade. The wartime complement of such a brigade is a headquarters company (also known as the "anti-VIP" company), three to five strike battalions, a signals company, and supporting elements.

On paper, there are sixteen such brigades. In peacetime, only the headquarters company and some of the battalion's companies are likely to be maintained at full strength. The two or three "regiments" operated directly by the GRU Central Apparatus are considered to be fully manned and operational as well.

There are also four naval *spetsnaz* brigades, each composed of a professional headquarters company, a midget submarine group, several Scuba and scout swimmer battalions, a parachute battalion, a signals company, and other support units.

The KGB also has its own units trained for unconventional warfare and special operations. These report to Department 8 of the 1st Chief Directorate. There has long been a rivalry between the KGB and the GRU in intelligence operations in general. It is probable that this condition applies to *spetsnaz*-type operations as well, since GRU Central Apparatus–controlled operations extend beyond the immediate theater level in scope and include many of the political targets that might more logically be considered the province of the KGB.

Experience suggests that the KGB's elements are oriented toward peacetime political-military and political-economic objectives, while the GRU's are directed more specifically against adversary war-fighting capabilities. Where the distinctions are blurred, teaming occurs (as in Afghanistan) or parallel operations with the same goal develop, with potentially counterproductive results.

GRU ground force *spetsnaz* units are tasked with a number of missions:

> Assassination of enemy political and military leaders during periods just prior to or immediately upon initiation of hostilities or other armed action. The killing of Afghanistan's President Amin and some of his principal loyalist officers during the December 1979 coup is an example. There have also been persistent reports of several attempts to ambush Ahmed Shah Massoud, one of the more successful Afghan rebel leaders.
>
> Destruction of nuclear weapons sites, airfields, command and control centers, ammunition dumps, fuel storage and pipeline facilities, and power generating and transmission structures. *Spetsnaz* brigades targeted against NATO utilize mockups of such installations at their training facilities, and fairly good ones at that; specific distinct weapons launch configurations for Lance, Pershing, and ground-launched cruise missile systems have been identified.
>
> Strategic reconnaissance and target designation. Like their U.S. counterparts, Soviet SOF mission assignments reflect a distinction between strategic reconnaissance and deep reconnaissance. In both armies, the latter is a tactical function assigned to echelons at or below the army corps level and limited generally to 100 kilometers from the tactical front. In practice, this distinction becomes secondary to the importance and sensitivity of a particular mission. Target designation of critical facilities for air strikes and stay-behind reporting of enemy movements are missions the *spetsnaz* share with many of their Western counterparts.

Support for airmobile or airborne operations. In addition to serving as covert advance party elements to facilitate power projection operations such as the events in Prague and Kabul, *spetsnaz* elements can infiltrate enemy-held territory and facilitate the aerial ambush (spotting for troops landed by helicopter on both sides of an enemy force caught in a defile). The Soviets are using this technique to good advantage in Afghanistan. The *spetsnaz* are also (ideally) capable of dressing and acting as enemy troops (as did the Germans in World War II with Skorzeny's unit during the Ardennes offensive and the Brandenburger Regiment's operations against the Soviets) to help disorganize and deflect potential enemy preparations for, or reactions to, large-scale airborne or airmobile raiding operations or attempts to seize and hold critical terrain during an offensive. One suspects, however, that in the conscript Soviet army, maintaining the requisite accentless language fluency on the scale required is no mean challenge.

Organization of stay-behind forces for partisan operations. This appears to be as close to a guerrilla warfare doctrine as the GRU units come, though there are indications of some teaming with the KGB in working with actual indigenous resistance elements. Despite the Soviet Union's support of "wars of national liberation" and proxy conflicts, the actual involvement of military *spetsnaz* elements in genuine guerrilla movements is the exception, not the rule. (They have been involved, with notable lack of success, in counterguerrilla advisory roles in Angola and Ethiopia.) This phenomenon should not be surprising. The Soviet experience with partisan warfare is of civil war and World War II texture; in both conflicts, cut off or surrounded military units merged with other elements, all Russian, to fight a conventional enemy on Russian soil. Thus, those who see a partisan support mission for the *spetsnaz* and those who believe the unconventional warfare dimension is not in the *spetsnaz* mission profile are both correct.

Spetsnaz elements are indeed expected to organize and carry out partisan warfare behind enemy lines in a major conflict. Those partisans, however, would be airborne-airmobile troops whose initial insertion purpose had been served and who could not be usefully extracted. The Soviets did just that following most of their airborne operations in World War II. They do not appear to be under the impression that there will be a great many potential guerrilla recruits in a "liberated" Western Europe.

Naval *spetsnaz* units are tasked with conducting reconnaissance of potential landing beaches, enemy naval facilities, and submarine approaches to fleet anchorages, as well as amphibious raids. The minisubs that created such a furor off the Swedish coast in recent years are of the type found in *spetsnaz* units.

GRU *spetsnaz* units are armed with the various Kalashnikov rifle models designed for use by airborne-airmobile units, specifically the AKM, the AK-74, and the relatively new short-barreled AKR carbine. A long-range HF radio, the R-392 or R-393, capable of AM voice, CW and encrypted burst transmissions, and a variety of explosives would also be found in most operational elements. Depending on the mission, one might additionally expect to find Strela hand-held surface-to-air missiles, silenced pistols, Dragunov sniper rifles, and grenade launchers.

Principal mission-area delivery means are (for the nonnaval units) clandestine prehostilities infiltration (certainly the preferred means); static line, high altitude–low opening, and high altitude–high opening parachute operations via the low-level-capable Antonov-12 or the larger Ilyushin jet transport; and the long-range versions of the M1-6 and M1-26 helicopters. Other military transport aircraft are available for longer-range missions, as is Aeroflot's civilian fleet. However, in any general conflict, the former are likely to be occupied with loading up the regular airborne divisions, while the latter must adhere to flight paths not necessarily amenable to more than limited *spetsnaz* insertion activities.

Naval *spetsnaz* paratroop elements would stage by airdrop in the same manner as the ground elements. They might also infiltrate via the naval air variants of the M-24 helicopter. Other elements would most likely infiltrate via small boat, Scuba, or minisub from a Hotel- or India-class mother ship submarine. The minisub groups are known to have tracked minisubs, which can trundle along a harbor floor (or a fjord, as Sweden is only too well aware).

Training for all elements concentrates on infiltration techniques, use of foreign weapons, mastery of target country languages and culture, survival skills, sabotage, silent kill techniques, target location and designation, and accurate reconnaissance reporting. Realistic raiding exercises are conducted against MVD (internal security) troops and forces guarding sensitive military installations. Such exercises are, of course, mutually beneficial in terms of lessons learned for aggressor and defender. However, the troops on the receiving end may have a different perspective, since the *spetsnaz* do not sacrifice realism for safety and fatalities are not unknown.

Like regular Soviet forces, *spetsnaz* units have been routinely cycled through the best live-fire training center the Soviets have, Afghanistan, regardless of nominal unit orientation, to spread the experiential wealth. For this and other reasons, the Soviets face some of the same sustenance and stability challenges as their Western counterparts. Nonetheless, they remain a considerable peacetime power projection asset and wartime force multiplier for a nation whose military philosophy stresses surprise, deception, and disruption as handmaidens to the offensive.

4
The Role of Special Purpose Forces in Eastern Europe

While little has been written of the non-Soviet SOF of the Warsaw Pact, they must be considered significant factors in the overall Soviet *desant* (disruption and distraction) effort in a general war and as occasional actors in Soviet Third World adventures. Particularly in a scenario where the Soviets opt for strategic surprise against NATO over a massive but obvious buildup, by initially committing only the available Soviet and Pact forces in East Germany, Poland, Czechoslovakia, and the western Soviet Union, the special purpose units of the Eastern European states are critical as force multipliers, and several have onset-of-hostilities missions.

At the same time, the Soviets are not blind to the potential for some units, particularly those of Poland and Czechoslovakia, to place nationalism before Communist brotherhood should internal political violence erupt, requiring suppression. Thus missions assigned to those forces (both conventional and special purpose) play on latent fears of a rearmed Germany, provide for taking on necessary but thorny targets that the Soviets themselves will bypass, and lead to the inescapable conclusion that the Soviets intend that those countries' best troops will have every opportunity to give their lives for the larger good of the Warsaw Pact.

Consider Poland's SOF. Their principal mission appears to be assisting their parent units, the 6th Pomeranian Air Assault Division and the 7th Naval Assault Landing Division, in the seizure of the Baltic approaches with but token assistance from the Soviets. The SOF involved here include the independent 3rd Sabotage and Reconnaissance Regiment (battalion sized) on the army side and combat swimmer detachments from the naval infantry. The 3rd Sabotage and Reconnaissance trains in demolitions, clandestine operations, and ground and amphibious reconnaissance, and its language training is almost entirely in German and Danish. More immediate support for the regular paratroop battalions going for north German and Danish strongpoints comes from each battalion's special diversion and sabotage groups, which are trained in sabotage and ambush techniques, and high altitude, low opening parachute operations. Unlike these units, which are directly under the 6th Pomeranian's command, the 3rd Sabotage and Reconnaissance reports to the intelligence directorate of the General Staff. The combat swimmers have training and equipment similar to their Soviet naval *spetsnaz* counterparts but a more focused mission: assisting in the seizure of the Skagerrak between Denmark and Norway and disrupting military traffic between the Jutland Peninsula and the Danish islands.

To continue in the war-gaming vein, we might expect the Soviets to throw most of their weight opposite the Americans and Germans in the central sector along the shortest avenue, Fulda-Würzburg-Frankfurt. From their standpoint, they can then circle down from the north, closing in on the Americans and Germans in Bavaria, and, they hope, inducing the French in Baden to opt for home and neutrality. This can succeed only in the "forces-available" scenario by getting a strong effort out of the Czech army as well as the handful of Soviet divisions in or immediately near Czechoslovakia in fixing German and U.S. forces in Bavaria. Toward that end, the Czechs' 22nd Airborne Regiment is oriented toward targets in that particular region. The regiment's 4th Battalion is designated a special forces unit and trains in the Bavarian dialect of German. Its skills include demolitions and other forms of sabotage, target designation, and sniping. Should the 22nd Airborne Regiment be assigned the (near-suicidal) task of cutting the lines of communication into and out of either Nuremburg or Munich, the 4th Battalion would probably have the mission of organizing the survivors into partisan groups.

One detects a different approach by the Soviets to the more politically reliable East Germans, whose Ministry for State Security is known to have thousands of active and sleeper agents in the West. It is also known to have an almost legendary ability to forge Federal Republic of Germany documents. Hence, unlike anywhere else in Western Europe, there is a significant potential for partisan-like disruption operations in addition to a large intelligence net.

To help the Soviets exploit the rear-area-chaos possibilities of this situation, East Germany had dedicated its 40th Parachute Battalion (its latest numerical designation; it has had several others), usually referred to as the Willi Sanger battalion. It trains in West German dialects and the American dialect of English. It also trains in U.S. Army doctrine and the use of U.S. uniforms, equipment, and weapons. It should not be surprising that they are available; after all, the East Germans have maintained a significant contingent with the Soviet mission to Vietnam for some time. In addition to the Willi Sanger battalion, the East Germans maintain combat swimmer detachments with mission and training similar to that of the Poles.

Hungarian, Rumanian, and Bulgarian SOF are minimal and are tasked only with supporting their national armies in their not-very-offensive roles of intimidating Austria, Yugoslavia, and Greece and perhaps attempting to seize Turkish Thrace. Weapons and equipment are little better than those of the regular armies, which are themselves second and third rate. This is not to say they do not play the roles in the grand Warsaw Pact design that the Soviets desire, for they surely do. They are well armed and trained enough to assist in their nation's limited missions against NATO and to intervene if required to quell civil disorder in a neighboring country whose best troops are, alive or dead, in the West.

Special Operations!
The combat skills are important . . .

. . . and the free world's SOF share them.

Getting there is also critical . . .

. . . whether by parachute . . .

Facing page:
. . . by helicopter . . .

. . . on the water . . .

. . . or under it.

Virtually any environment must be "home."

PSYOPS also play an important role, and in the United States some unique means are available to deliver the message.

5
Low-Intensity Conflict Approaches

Special operations and *low-intensity conflict* are often used interchangeably—perhaps as inevitable as it is incorrect. Low-intensity conflict encompasses a wide range of preventive and preemptive measures, as well as those appropriate to open hostilities, and a range of conflict forms—support for resistance, counterinsurgency, foreign internal defense, psychological operations, counterterrorism—which certainly can include but are not limited to special operations per se and the use of dedicated SOF.

That said, whatever other elements are utilized in low-intensity conflict, SOF inevitably play a critical role. The nature of the conflict itself and the goals usually require, at least for the West, application of something less than total combat power yet more complex than mere provision of security assistance supplies and verbal support. For the Soviets, SOF are also critical to counterinsurgency efforts, though their use in Afghanistan is not qualitatively different from the *desant* and target destination roles they have in more general war-fighting scenarios.

It is appropriate now to look at some of the larger environments in which not only SOF, but the national role perceptions that provide their existence rationale, are operating. This examination begins with Soviet approaches to Marxist counterinsurgency in comparison with doctrines prevalent in the West. France's low-intensity conflict stage, Africa, and its unique approach to commitments there were addressed sufficiently in chapter 2. The other two Western powers with significant ongoing low-intensity conflict situations are the United States, with its extensive (and extended) commitments in Central America, and the United Kingdom, with its troublesome living legacy of medieval history, Northern Ireland. In these latter, SOF are critical components, but just that, in an overall low-intensity conflict strategy. How well those strategies are being articulated, coordinated, and implemented on the ground will be the focus here.

Western Low-Intensity Conflict Approaches versus the Soviet Union's Marxist Counterinsurgencies

Increased Soviet military involvement outside the Warsaw Pact—not just in advisory or assistance roles but in actual combat operations—is a fact of modern life. Soviet military planners have, like their Western counterparts, found themselves obliged to lay aside their preoccupations with general conflict scenarios to deal with low-intensity exigencies of their own.

Many aspects of low-intensity operations are of general applicability. They are usually considered subnational (guerrilla, counterguerrilla, counterterror, and so on); if international, they are fought with limited forces for limited goals, generally without manpower or industrial base mobilization. They are carried out to preserve or create a political, military, and social environment favorable to a sponsoring power's national security and foreign policy goals, as well as to further the interests of client states or the progress of a resistance movement. Where the major differences between Western and Soviet low-intensity operations lie are in the style of execution and the articulation of national security goals.

Wars of National Liberation

From the Western perspective, Soviet military involvement with revolutionary groups is very limited; the military appears in the role of client-state support against other revolutionary or resistance groups. In the Soviet view, the military supports a successful revolutionary client in the consolidation phase against counterrevolutionaries; the assumption of power after a guerrilla campaign or a coup is seen not as an end but as a beginning. A revolution is not just engineered; it must be maintained. This line of thought is based on the Soviets' own historical experience in the Russian Revolution and ensuing civil war and must not be dismissed as empty rhetoric or transparent attempts to justify support for puppet governments.

At the operational level, however, actual military involvement in the resistance or guerrilla phase is indeed rare; that phase normally is the province of the KGB.

Relationships with Host Country Forces

The current Western approach to conducting stability operations in support of a client state is to provide military and economic aid and program guidance and conduct advice and assistance activities with small cadres of noncombatant officers and noncommissioned officers. If a sponsoring power's own troops must be employed against indigenous guerrillas, the effort will appear to have been taken over from the host country. In the case of external aggression, use of sponsoring power troops as a shield is viewed as acceptable. However, command and control of host country forces is still vested in its own commanders, who may or may not accept as much guidance as outside advisers want to give.

By contrast, the Soviets insist on control commensurate with their level of involvement. In Angola, the initial logistics buildup and deployment of Cubans that led to the victory of the Soviet-backed MPLA over the FNLA and UNITA in 1976 was controlled entirely by Soviet officers. Similarly, a Soviet military command-and-control structure presided over the 1977 equipment airlift and subsequent direction of Cuban and Ethiopian forces against Somali-backed rebels and invading regular Somali forces in the Ogaden. In both cases, the Soviet-engineered victories have not been as complete as they once seemed.

In Angola, UNITA has become a formidable threat to the MPLA and Cubans, both of which have suffered major setbacks at the hands of UNITA and South African forces. In Ethiopia, Cuban forces are increasingly unhappy with the choice of fighting Eritrean and Tigre rebels or relative inactivity in the Ogaden. In both cases, the Soviets no longer play an active, direct military role, though their clients are still dependent on them for aid and technical assistance.

In Afghanistan, the Soviets initially attempted to maintain the fiction that their own troops conducted only limited offensive operations in support of indigenous Afghan forces. However, the Soviets alienated much of the leadership core that remained in the Afghan Army in 1980 with a brutal, blatant assumption of all command functions. Actions ranging from simple but widespread arrogance to public execution and, in one case, the annihilation of a unit that would not take orders from Soviets have characterized the Soviet approach to stiffening the Afghan army. Consequently there have been few Afghan Army–led operations, and what there have been cannot be considered combat effective. The Soviets, therefore, while they still pushed unenthusiastic Afghan units ahead of them as tokens of unified operations, assumed the burden of organizing and backing up all offensive tactical operations, which they retained throughout their active role in the war.

Relationships with Host Country Population

The idealized Western approach to counterguerrilla operations, not always achieved in practice, is to secure populated areas and then to provide a shield in the form of continued military operations or static protection while various psychological, economic, medical, educations, and other programs are implemented to co-opt the population and diminish the guerrillas' appeal. The recognized hazards of this approach are that there are usually insufficient forces to secure all rural populated areas and that guerrillas can be fought only when they operate away from their population support bases.

The Soviets, by contrast, appear to have only limited enthusiasm for converting populations that support guerrillas. While their first two years in Afghanistan reflected an attempt to separate the "fish" (guerrillas) from the "sea" (the populated rural valleys), it failed as much because of small unit tactical ineptitude and lack of enthusiasm on the part of the Soviet troops as it did because of popular anti-Soviet sentiments. Through fear, carelessness, or frustration, Soviet leaders began more and more to bomb and shoot unarmed villagers indiscriminately during sweeps.

However, it became apparent, as the Soviet presence wore on, that their answer to population and resource control is systematic eradication of the rural agrarian infrastructure. Villages were leveled, crops and livestock destroyed. The only protected population were the client's adherents, who are now clustered in the larger cities. Only the proximity of the Soviet Union, with its technical support, and the persistence of rural-urban trade despite official prohibitions against it, kept life going for the protected urban population.

Special Operations

Special operations are, for the West, a large part of what comprise low-intensity operations. They include foreign internal defense training and advisory efforts, aid to resistance elements, unilateral raid, sabotage, and counterterror operations and psychological warfare.

For the Soviets, the special operations role in its low-intensity conflicts has included some of the same elements and others as well. Certainly the Soviets in Afghanistan have employed *spetsnaz* elements in their reconnaissance, raid, and target designation roles. They have also used these units in assassination attempts against *mujahedin* leaders. However, the *spetsnaz* units are not known to have been involved with any bona-fide resistance-support effort or to counter rebel forces with *spetsnaz*-directed indigenous forces similar to the U.S.–led Montagnard units in Vietnam; at least, they have not since 1979, when Afghan Baluchi tribesmen that *spetsnaz* teams were successfully turning against the *mujahedin* themselves turned on the unsuspecting and uninformed Russians in retaliation for the December intervention. There are some reports of *spetsnaz* taking on *mujahedin* or Baluchi disguises, but while the latter may have some relation to actual attempts to divide and disrupt support lines, the *mujahedin* charades are not destabilization but deception, similar to that attempted by Skorzeny's troops dropping at Ardennes dressed as Americans during World War II's Battle of the Bulge.

Tactics and Techniques

The Soviet approach to low-intensity operations postulates a revolutionary client who is being supported against counterrevolutionary, reactionary forces. It also features terrorizing populations who do not readily submit. As its operations in Angola and Ethiopia in 1976–1977 involved command, control, and logistical support of an essentially conventional army in open combat against inferior forces, the Soviet army's existing operational methods appeared to have served adequately.

In preinvasion Afghanistan, the Soviets were for the first time involved in a major advisory effort involving a client army's operations against a truly popular, apolitical resistance. At the time, the Soviets had nothing in the way of doctrine and little counterguerrilla tactical acumen to offer. The situation was tolerable only because there were at the time some good Afghan Army units and because the diversity of nationalist and Islamic fundamentalist drives that gave the *mujahedin* life also set many of the groups at each other's throats. With the coalescing and growth of the resistance and the disintegration of the Afghan army, both caused by the Soviet occupation, the Soviet Army employed a number of measures with varying effectiveness in its search for workable counterinsurgency methods.

The Soviet Union's first conscious effort to defuse popular resentment in Afghanistan also reflected its underestimation of the combat requirements its forces would face. Outside of the quick strike *spetsnaz* and airborne units involved in the original seizure of Kabul and Bagram airfields and the overthrow of Afghan leader Hafizullah Amin, the divisions that drove overland into Afghanistan in December 1979 were Category III, composed largely of Tadjik and Uzbek reservists, Muslims who often could not communicate with their ethnic Russian officers. It was thought that their presence would somehow mollify the Afghans; it was certainly not intended that these troops would have to fight. As it turned out, only the latter case could be said to have held true: the Muslim reservists also did not expect to fight. Some of them, according to Afghan sources and other reliable reports, engaged in open fraternization, leaving weapons and food for the *mujahedin*.

Those troops that did undertake operations did so without enthusiasm and were soon seen by a number of Soviet senior commanders as little better than the Afghan Army. Some units refused en bloc to go on operations against their co-religionists and ethnic brothers, obviously a bad example for the other Soviet units and the Afghan Army alike. The 105th Guards Airborne Division and separate airmobile brigades became the only truly reliable Soviet units and were stretched thin. When the Muslim reservists were sent home in mid-1980, they were replaced by ethnically European troops.

The enthusiasm of the conscripts in the Soviet motorized rifle divisions still in Afghanistan appears to most observers to be little better than that of their Turkic predecessors, and their ability to fight light in the mountains or react aggressively to convoy ambushes is not highly regarded. However, their are indications that the Soviets learned to make positive use of this negative trait during times of more active combat. Ground formations have been sent along prime ambush routes, made contact, taken casualties, and suffered vehicle losses, all to set the *mujahedin* up for Soviet helicopter gunship and close-air support strikes.

The Soviets have found, as did the U.S. forces in Vietnam, that helicopters and other aircraft are indispensable for counterinsurgency warfare. This discovery was not gradual; in fact it coincided with the mid-1980 withdrawal of the Turkic reservists and the simultaneous appreciation that virtually all of the vehicular lines of communication and supply ran through what had to be considered hostile territory. Accordingly a rapid buildup was effected to reach what are thought to be the peak Soviet general strength ranges of 250 to 300 helicopters (at least 100 of which serve either primarily or secondarily as gunships) and over 300 fighter-bombers of various types, in addition to fighter and long-range bomber assets available in the Turkestan Military District of the Soviet Union.

The Afghan army and air force have additional helicopters and fighters. These aerial assets have become the principal strike weapons of the war, rather than artillery and tanks, as well as the principal resupply means for isolated fortifications. One tactic used with some success has been the identification of resistance groups by far-ranging *spetsnaz* elements, followed by landings of airborne or airmobile brigade troopers to drive the *mujahedin* into more open areas, where they and any nearby civilian areas are struck by helicopter gunships and tactical aircraft, followed finally by mop-up operations conducted by armored ground units moving up the valleys. The Soviets, however, have lost over 300 aircraft and over 5,000 vehicles in the war so far and apparently suffered heavily in these offensives; certainly the Soviets did not remain in the secured areas, and ultimately the *mujahedin* moved back in.

What the Soviets have achieved is population control in a very grim sense. Between village bombings, documented nerve and blood chemical agent attacks, destruction of livestock and crops, and booby-trapped toys, fountain pens, and dolls dropped by helicopter, the potential "sea" for the guerrilla "fish" has been either corralled in the cities or removed to Pakistan, if not eliminated where they were found. This has resulted in diminution of ready personnel and sustenance assets for the guerrilla units and thus limited their dispersal options as well as their operational range. On the other hand, it has also eliminated the guerrillas' concerns for the safety of families and brought more dispossessed peasants into *mujahedin* ranks. In effect, it has created vast free-fire zones for the Soviets and increased recruiting for the guerrillas. The proximity of China, Pakistan, and Iranian Baluchistan reduces the guerrillas' supply and safe-haven concerns.

Conclusions

On balance, the Soviet ability to successfully develop and manage military programs to sustain political events and personalities established by the KGB is considerable, as long as clients are willing to cede command and control functions and the adversary offers battle under more or less open conditions, without the training or equipment to require more than Cuban proxy tactical responses.

Soviet politicopsychological ability to convert hostile population is nil; the preferred tactic is to force the problem population into exile or otherwise eliminate it. Small unit counterguerrilla tactical doctrine for the Soviet Army proper has been generally lacking because of inexperience but also because of conscious training emphasis on centralized command direction.

A counterguerrilla war is a platoon and company commander's war, but the only Soviet officers encouraged to think in terms of independent operations are the full-time airborne, airmobile, and *spetsnaz* unit officers, whose units are feared and respected in Afghanistan. The motorized rifle and tank unit soldiers are not. Thus, while the Soviet Army undoubtedly gained valuable combat experience at all levels in Afghanistan, it found that it was not winning the war there by any measurable standard, unless slow extermination of rural life is considered winning. Overall the Soviet Union's ability to engineer a coup or conduct a speedy, country-crippling intervention is impressive, as is its ability to translate disinformation into useful destabilization programs. But its ability to combat a determined, popular rebellion outside the Soviet Union is decidedly limited. It is a prisoner of its failure to develop a coherent counterguerrilla tactical doctrine in general and its own dogma in particular, a dogma that simply accepts an adversary relationship with genuine popular resistance rather than making any concerted attempt to co-opt it.

The United States in Central America

In the Iran-contra congressional hearings and the media-solicited commentary that saturated the airwaves during and after them, people in the United States and the rest of the world got glimpses of many perspectives about Central America. U.S. and Central American public officials, media commentators and other observers from all hues of the political spectrum, Nicaraguan Sandinista and contra leaders, and ordinary citizens served to bring various aspects of and views relative to a region in conflict to the world's attention. The coverage itself and the positions stated (or extolled, or explained), regardless of the diversity of them, pointed up two major facts of international political life: that the United States takes special (one might say proprietary) interest in Latin America, particularly the Central American states, and that that interest and the forms in which it is manifested have tremendous impact on the people and governments of the region.

Neither of these conditions is new. U.S. special interest and acts of state bearing out that interest go back over 150 years. The level of intensity of that interest has, of course, waxed and waned as wars, crises, and confrontations overseas captured attention and resources from time to time. Still, since the enunciation of the Monroe Doctrine, the United States has (without ever being so callous as to suggest it in so many words) regarded Latin America as its own sphere of influence. Until around the middle of this century, the United States saw nothing wrong with outright intervention when it felt its interests, either direct or as regional suzerain, were threatened. Contrary to most current liberal cant, however, interventionism as a policy now has no serious adherents, although interventions have been carried out for brief periods of time (Dominican Republic in 1965 and Grenada in 1983) under the auspices of regional security organizations. But it is not necessary to intervene unilaterally in order to exercise influence in pursuit of security objectives. Engineering and/or supporting covert operations or a coup against a hostile regime is one means. That the United States has utilized and does utilize such means is a matter of public record, both defended and attacked, but that is not the subject here. Another means (which is the subject) is the security assistance and training component to U.S. foreign aid programs for the region.

Regional Approach

U.S. security assistance and training programs targeted against externally sponsored insurgencies in Latin America have been going on for decades. If anything, the most recent manifestations of emphasis are a reprise of the more effective aspects of the policies of the Kennedy-Johnson years on combating Communist-inspired expansion wherever it was found. To be sure, they are a long way from the "bear any burden, pay any price" counterinsurgency assistance efforts of the 1960s in terms of funding and numbers of trainers and advisers on the ground. The only areas in which current programs outpace those of twenty years ago are in the impact of technology, and visibility to a "Vietnam-syndromized" media.

One other similarity between current programs and those of the Kennedy-Johnson years is that the principal implementers of advisory and training programs are small teams drawn from U.S. SOF units or individuals with special operations backgrounds and qualifications. A great many other elements provide direct or indirect support, particularly in Central America, where activities in several countries are derived from and relate to a U.S. strategy for the region as a whole and are backed up by application of airborne and stationary intelligence and target acquisition platforms, large-scale logistics infrastructure development, and covert (or more appropriately, formerly covert) operations. But the common thread throughout the direct U.S. assistance efforts is the promotion of tactical skills and grass-roots psychological and political efforts to enable a host country to defeat an insurgency, both on the battlefield and in the popular "hearts and minds." For the most part, the implementers of these efforts have been U.S. SOF elements or SOF-qualified individuals. Why SOF personnel have long been involved, and today have a preeminent role, in what has for the United States in effect become a regional low-intensity conflict campaign goes to the heart of both U.S. strategy in the region and recognition by the United States of domestic and regional political realities.

In fact, much of the policy-driving rationale for prioritizing U.S. special operations programs articulated by the secretary of defense in his FY 85 Annual Report to Congress could have been written with the special operations role in Latin America in mind, particularly its emphasis on countering destabilization efforts and "reducing the likelihood that U.S. forces will become involved in combat."

In most of the Third World, and certainly in Latin America, former secretary Weinberger's expressed preference for security assistance, training, and advisory efforts over direct conventional force intervention is more than matched by similar preferences of the countries involved. And quite apart from the onerous symbolism that would be attached to conventional intervention (not to be confused with quick, "clean" operations like "Urgent Fury" in Grenada) in the eyes of the American public, there are sound doctrinal reasons for avoiding commitment of U.S. ground combat forces. If a sponsor (in this case, the United States) must employ its own troops against guerrillas indigenous to another country, then the counterinsurgency effort will appear to have been preempted from that country's government. The resistance movement can then denounce the host country government as "imperialist puppets" and redirect its own appeal to the populace as one of driving out "foreign invaders."

Army Special Forces and PSYOPS-CA personnel and navy SEALS are trained to be trainers as well as operators, and the array of skills that would be applicable to their missions of carrying out guerrilla warfare behind enemy lines is also applicable to the counterinsurgency mission. These U.S. SOF have for some time been involved in training conventional Latin American forces in small unit counterinsurgency tactics, as well as providing more specialized training for reconnaissance and quick-reaction units. They have also carefully, through PSYOPS instruction and the examples set by CA unit deployments, attempted to alter some traditional military and political establishment views of the peasantry. Why not just use personnel from conventional U.S. units? In some cases that does indeed occur. In specialized areas of wide-net communications, technical intelligence collection, engineering, schools curriculum development, and some others, assets beyond those normally available to SOF are required. The U.S. Southern Command, which implements and coordinates all U.S. security assistance and other military activities throughout Latin America, draws on the in-place U.S. infantry brigade in Panama and resources in the United States for many requirements. Indeed that brigade has been reorganized as Task Force Bayonet, specifically to serve the security assistance and low-intensity mission.

For basic counterinsurgency training, however, special operations personnel make the most sense, not only because they have received training in counterinsurgency but because they have been trained as well to organize and assist resistance movements in their own right. Knowing what the insurgent will be trying to accomplish, both militarily and psychologically, the task-organized, multicapable SOF training team can train at the appropriate levels in the tactical, psychological, and local political measures that can preempt the insurgent.

The degree to which that actually happens is greatly dependent not only on the skills base that the SOF personnel bring with them but also the receptivity of the host country. Established local attitudes regarding warfare, how the peasantry should be treated, and how to integrate military operations into an overall strategy that has a realistic chance of finding and then crushing guerrilla units in the field and also preventing their regrowth through preemption of the population are not easy to alter. Moreover, every political entity in Latin America that receives significant U.S. support is sensitive to being seen as too dependent on that support.

Thus, program manifestations vary greatly. The United States has been ingenious, as well as persistent, in adapting to the realities it cannot change while pushing for the changes that have been possible. Those realities, however, born of the region's traditional relationships between the military and the masses, have had to be factored into virtually every U.S. assistance program activity.

Regional Operating Environment

Standing national defense establishments are an outgrowth of evolved perceptions that a state requires national protection and an armed tool of foreign policy beyond that afforded by militias or as-required levies on the population. Such establishments are perceived as principally being oriented outwardly (that is, against potential foes external to the nation's boundaries). Certainly this is the European-American tradition. That tradition does not deny military roles in maintenance of civil order, and in fact most Western countries (including Canada) have utilized their armed forces in suppressing actual or incipient insurrections, intervening in civil disorders, and otherwise backing up the civil power. For the most part, however (the British operations in Northern Ireland are a notable ongoing exception), Western military involvement in domestic internal security operations is currently focused on providing specific or backup support for counterterror programs.

In Latin America as well, national military establishments were also initially oriented against external threats. They grew out of the forces raised first to combat those of the Spanish crown and then, as the victors' successors found issue with postcolonial border adjudications, to fight each other. The Chaco Wars of 1932–1935 in which much of South America was aligned against Paraguay, and the War of the Pacific in 1879 in which Chile defeated the combined forces of Bolivia and Peru, were regionally significant bloodlettings. Border disputes over the Beagle Channel and the Andean demarcation line still simmer between Chile and Argentina, as do the results of the War of the Pacific between Chile and its former enemies. Honduras and El Salvador fought a brief war in 1967 over border-related differences; the war ended not because the disputes were resolved but because both sides ran out of spare parts and ammunition. Also unresolved are the issues that led to the most recent conventional war in the region: the Falklands-Malvinas conflict between Argentina and Great Britain in 1982.

Although Latin American military establishments can indeed trace their origins to the same rationales that gave birth to their European counterparts, it is also true that a variety of factors have given them a much more distinctively internal security character than those of Europe or North America. One factor has been the military's institutional view of itself as the guarantor of national sovereignty and identity. Whether this view is cynically or altruistically held, the milieu of upper-class factionalism and unmet peasant and worker aspirations in which it operates virtually ensures that the result is the same: a long history of military involvement in (or control of) the domestic political process.

Another factor has been that, aside from those instances of international conflict like those already noted, the principal adversaries that Latin American military establishments have faced have been their own countrymen, in the form of riots, mass general strikes, armed insurrections, revolutionary guerrilla forces, and even rebel military units backing rival politicians or generals. Such an environment provides unique pressures for the process of manning, training, and equipping Latin American military establishments. Traditional border disputes and concepts of national prestige give impetus for creating formations with adequate mass and sophistication to protect territorial integrity and project a credible deterrent to antagonistic neighbors. On the other hand, the need to defend the nation from internal threats and to impose martial control over civil functions—for whatever reason—makes counterinsurgency, PSYOPS, and, at times, law enforcement operations major factors for the orientation of forces.

There is a danger in ignoring distinct national differences in the course of making sweeping regional generalizations. The fact that the likeliest threat to most polities is domestic does not mean that the response should or will be the same from country to country. When the Tupamaro threat first gripped Uruguay in the early 1970s, for example, the government tried heroically to keep the military out of the fray. As the scope of the threat grew to a level where the military had to be called in, it was the people, through their elected legislature, who initiated the action, not the armed forces. In other countries, such as Argentina and El Salvador, military and police officials found common cause with powerful right-wing politicians to circumvent legal process with volunteer death squads. That this is virtually a thing of the past in both countries is tribute to evolving public awareness that such measures, beyond being morally reprehensible, tend to create more guerrillas and terrorists than they kill.

There are, however, some common features around which consideration of Latin American assistance operations can be structured. The first is that regardless of the original causes and dogmas, almost all current insurgencies in Latin America are perceived not only in the United States but in the countries themselves as receiving support in one form or another from Cuba, either directly or through Nicaragua. Some have also taken on a new character: involvement of the insurgents with the Latin America–to–U.S. drug traffic. As a consequence, the United States sees a multidimensional threat to both regional stability and U.S. national security and is therefore deeply involved. The United States recognizes and tries to stay aloof from but occasionally finds itself the victim of the remaining country-to-country differences (as occurred when Honduras, incensed over the high proportion of Salvadoran battalions trained at the Regional Training Center established by the United States, forced its closure in 1985). However, it concentrates on what it considers appropriate common regional responses to a common regional problem.

The apotheosis of that problem, in U.S. eyes, is Nicaragua. The Sandinistas' military buildup makes Nicaragua a credible conventional threat to its neighbors, whether manifested in outright aggression or in pursuit of contras, and support for guerrilla organizations in El Salvador has long been funneled through Nicaragua. But while Nicaragua has been the principal bête noire for the Reagan administration in Latin America, Marxist guerrilla threats elsewhere are seen as similar, if currently slightly less threatening, expressions of resurgent Cuban determination to destabilize the region. In the 1960s, U.S. aid and military advisers played a major role in bringing a number of Cuban-instigated insurrections to an end. The Reagan administration saw the contest developing anew, with its even more insidious drug trade implications, and has committed itself heavily to countering the Cuban-Nicaraguan threat through support for a full range of internal defense measures in the most affected countries. By way of illustration, two of the more heavily involved client states and those elements of U.S. country-specific and regional strategy being applied in them warrant a closer look: Honduras and El Salvador.

Honduras

Honduras enjoys the dubious distinction of having both of Central America's major current conflicts taking place along—and sometimes within—its borders. Remote areas of its frontier with El Salvador are frequently utilized as base camps and logistical infiltration routes by the Marxist guerrillas fighting in that country, and equally remote areas not far from the Honduran-Nicaraguan border serve as training and supply bases for the Nicaraguan contras in their campaign against the Sandinistas.

The United States considers both conflicts to be part of a regional campaign and views Honduras from two perspectives: (1) Honduras must be protected from Sandinista-Cuban aggression, either overt or in the form of a sponsored insurgency similar to El Salvador's, and (2) because of its necessarily anti-Communist stance and its location on the borders of existing conflicts, Honduras serves as an ideal base for supporting U.S. (and U.S. clients') activities.

Honduras cannot ignore U.S. interests, even when their manifestations in the form of frequent major exercises (which justify maintenance of support bases), reconnaissance flights in support of the Salvadorans, and the contra training camps may exceed Honduras' anti-Communist zeal. As the principal mainstay of Honduras' economic and military well-being, the United States enjoys a certain amount of latitude in terms of pursuing its regional security goals.

There are limits, however, and while it is an article of faith in the Western liberal community that Honduras submits to the United States to garner those essential aid dollars, reality is something else entirely. For one thing, Honduras does not give the United States a blank check. It closed down the U.S.-established and manned Regional Training Center several years back when it decided the Honduran Army was getting less benefit from it than the Salvadorans. Rivalry with the Salvadorans, dating from the last century and exacerbated in the 1967 "soccer war," also has at times affected the alacrity with which interdiction operations have been run against Salvadoran guerrilla infiltration routes. More recently, the government has quietly told the United States to ease off on the major exercises and has publicly stated that it does not want U.S. forces training contras within its borders.

Another consideration is that Honduras recognizes, in part because it has been aligned with the United States, that the Sandinista government does indeed pose a threat. There is disagreement within the Honduran military regarding how much contra aid activity to allow, how public to be on ostensible aid limitation activities (such as seizures of planeloads of equipment), and which military departments (not to mention individuals) will get what portions of such seizures or other assessments. There is concern that too aggressive a stance will provoke a general conflict; that concern is in fact what underlay the government's reluctance to move its forces into confrontation with the Sandinista incursion in early 1986. But the government knows that many of the seeds for a Salvador-like insurgency are in place, and while armed attacks against government symbols and personnel have been rare in recent years, the government received a real scare in 1983 when Nicaraguan-sponsored elements attempted to establish an armed guerrilla movement. The insurgents failed, but that event and the periodic skirmishing between Honduran elements and Sandinistas on the border keep the threat of further conflict real.

Within this murky swirl of conflicting internal and external pressures, the activities of SOF and other units assigned specific counterinsurgency tasks serve the demands of the moment. The major areas of consideration for our purposes here are performance of dedicated counterinsurgency and rapid intervention forces and capabilities and the quid pro quo with the United States.

Forces and Capabilities. Six army batallions are deployed along the borders. They cannot—and do not try to—cover each mile of frontier. The distances are too great, the terrain too rugged, and the likelihood of undesirable confrontation with Salvadoran guerrillas or Sandinistas too strong. However, U.S. experts report that while unit patrols may be thin on the ground, the cultivation of the local populace by unit intelligence officers has been extensive. Information thus gained is augmented by technically derived intelligence provided by the United States. When major interdiction operations do become necessary, the information base upon which they are launched is usually quite good (and better in many cases than in neighboring El Salvador). They also cannot provide much coverage in routine situations (other than joint operations with the Salvadorans or reacting to a fracas with the Sandinistas) because in each battalion, only one company is generally available for patroling or manning positions. Another company is usually in training, while the third is off duty.

Specific units that maintain more specialized capabilities are the

Airborne Battalion, located outside Tegucigalpa, the 2nd Battalion, located nearby, and the Special Squadron. The Airborne Battalion and the Special Squadron may be considered offsprings of the 2nd Battalion. The 2nd for a long time was the repository for airborne and counterinsurgency training and may still contribute instructors to the Airborne and Ranger Schools, which are not (at least nominally) controlled by the Airborne Battalion. The Airborne Battalion and the 2nd are not organized internally in much different fashion from the other (essentially light) infantry battalions of the army, with obvious exceptions in the Airborne Battalion's case in the area of parachutes and other air items. They do, however, receive the pick of new recruits and are particularly trained for airmobile (helicopter-borne) rapid intervention missions, as well as counter-guerrilla sweeps. Additionally, because of the U.S. penchant for frequent large exercises in Honduras, the Airborne Battalion can periodically conduct airborne training operations from U.S. Air Force airframes in numbers most other Latin American paratroop forces can only envy. While a battalion-sized parachute drop in a counterinsurgency operation may be difficult for some to rationalize, the Honduran Army's consideration of the Sandinistas as an overt, conventional threat makes the utility of a paratroop force more recognizable. Both battalions have trained with U.S. forces on maneuvers and as recipients of mobile training teams.

The air force maintains a squadron of U.S.-supplied A-37 Dragonflys (still the best counterinsurgency close air support platform) but has utilized them only sparingly. They do receive extensive training, however, during the regular U.S.-Honduran training exercises. The Coast Guard is miniscule but has been building its patrol and interdiction capabilities under U.S. Navy special warfare tutelage.

Honduras has had practical experience with an insurgency that contributes heavily to their training data base. In 1981, Nicaragua, with Cuban assistance and funding, recruited nearly 100 Honduran peasants under a deception that promised training in mechanics, agrarian reform, and financial support for volunteers' families. These recruits were given weapons, tactical, and demolition training throughout 1982, as well as extensive political indoctrination, in Cuba. They returned in 1983 as the armed element of the Central American Workers Revolutionary Party and received further "practical application" training fighting alongside Sandinista units against the contras.

In July 1983, they infiltrated into Olancho province under the party chief, Dr. Reyes Matta. They avoided patrols until reaching their intended base outside Nuevo Palestina. They suffered some internal dissension over the role of the church in revolutions and committed the egregious error of waiting until they had consumed their supplies before sending a team out to get more. Three days later, on August 1, three rebels deserted; two of them turned themselves in five days later, and the Honduran government reacted.

Elements of two battalions moved into blocking positions to prevent guerrilla entry into Nuevo Palestina or escape back to Nicaragua. In the meantime, PSYOPS and CA teams carried out a loudspeaker, poster, and face-to-face communications blitz with the civil populace and the enemy, providing protection for defectors and families, financial incentives for information from civilians and desertion by the guerrillas, and open broadcast of deserter statements. The results were a building picture of guerrilla locations, starvation among the guerrillas, and additional desertions. On August 11 the Special Squadron deployed small ambush teams close to suspected guerrilla locations. Six days later, the leader, Dr. Reyes Matta, was killed. In a month of operations, the Hondurans killed fifty-one guerrillas, took in twenty-four as prisoners or deserters, and found the bodies of eleven who had died of starvation. The incipient movement had been crushed by quick recognition of the problem; acting to seal it off and deny the guerrillas' access physically and psychologically to the populace; acquisition, analysis, and dissemination to users of information derived from the populace and deserters; and hunter-killer operations by small units of special forces armed with good intelligence. Those who wonder at the apparent foot dragging of the Honduran Army on the occasion of the Sandinista border crossing—which the Hondurans knew was not invasion but rather pursuit of contras, who in fact handled the Sandinista threat unaided—can look to the short-lived Olancho insurgency for a better understanding of what Honduras can do when it, as opposed to the United States, perceives a clear threat.

The Quid Pro Quo. Relations between U.S. SOF personnel and the Hondurans have been steady and close. U.S. Army Special Forces teams regularly conduct field exercises with Honduran units as part of the larger exercises. MTTs from the Special Forces battalion in Panama, as well as those from Fort Bragg, have worked with the Airborne Battalion and with the specialized schools. Air force special operations advisers and navy SEALS have provided MTTs for their counterparts too, and a U.S. SOF MTT conducted some of the Special Squadron's early counterterror training.

In addition, the once virtually nonstop stream of exercises designed to show the flag and build actual regional operations experience for a number of U.S. conventional units also provides several local long-term benefits to the security assistance program. First, the continuing nature of the exercises has necessitated a full-time U.S. support infrastructure that would otherwise have been almost impossible to establish as an overseas base or as a permanent augmentation to the U.S. military assistance effort operating in Tegucigalpa. That base supports operations in El Salvador as well as training for Honduras. Second, Honduran forces have a regular opportunity to conduct field training on a scale, and against sufficient numbers of "enemy," that would otherwise be impossible to duplicate. Third, the conventional forces deployed by the United States for such exercises have armament and equipment levels similar to that of the Sandinistas across the border and thus graphically, and continually, bring home to Honduras what they may have to face. And fourth, the rotating presence of various support elements does ensure the capability of rapid buildup for U.S. forces if required or of provision of rapid support for Honduran forces. This capability was illustrated when U.S. CH-47 Chinook helicopters were diverted from exercise tasks to lift Honduran troops into the area of the spring 1986 Sandinista "incursion."

U.S. SOF have not, however, thus far been employed in training contras in Honduras. The 1986 contra aid vote did provide for such training but could not (at least openly) mention where. Honduras (as well as Panama and El Salvador) has attempted publicly to distance itself from this step. In a covert war everyone knows about, Honduras can still turn the other way, but direct involvement of U.S. military personnel training insurgents on its soil escalates the issue, in the Honduran mind, possibly that one final notch that the Sandinistas will not tolerate. Under conditions of outright Sandinista invasion, Honduras would have no option but to do the next-to-last thing they want to do: turn to the United States for intervention. For lesser threats, U.S. assistance is necessary, and cooperation with the United States has been an acceptable price. But for lesser threats, it believes that its own counterinsurgency and direct intervention forces can and should take the lead role. Based on results to date, there is no reason from either the U.S. or Honduran viewpoints why they should not—or for any change to the good advice and assistance relationship with U.S. SOF.

El Salvador

The bulk of the U.S. efforts on behalf of El Salvador (and the United States' own concepts of regional stability) are political and economic (despite popular misconceptions to the contrary). However, neither of these elements can operate without at least effectively contesting, if not defeating, the rebels' armed strength.

U.S. Military Role. Principal aspects of U.S. assistance activities that may be considered counterinsurgency operations fall into three categories:

1. Training assistance by U.S. SOF and other elements at training centers in-country.
2. External training assistance by U.S. SOF and other elements (Honduras, Panama, and the United States).
3. Operations, Plans, and Training advisers with major commands.

Other activities that directly or indirectly support the internal defense and development mission, and/or U.S. regional strategic goals, may or may not, in a given instance, constitute support operations. Generally they fall outside the active military operations portion of a counterinsurgency support (or foreign internal defense) campaign.

Despite official protestations that the fifty-five-man advisory mission strength allowance is sufficient, the United States has gone to great lengths to augment its in-country training capabilities with extensive efforts outside El Salvador. Moreover, MTTs that stay for less than three months and nonmilitary specialists provide additional training assets.

Principal thrusts of the U.S. training effort have been to:

Wean the Salvadoran Army in general away from large unit "jungle bashing" through emphasis on small unit leadership, sustainability, and mobility.

Avoid alienating the populace through undisciplined application of ground or aerial firepower.

Take the war to the enemy through active patroling and denial of population centers.

Eliminate a "9-to-5" concept in military operations.

Forces and Capabilities. In addition to the regular infantry battalions and essentially defensive paramilitary security forces, El Salvador has organized forces with dedicated, offensive missions. They represent a conscious effort to adapt elements of their military structure to the specialized mission requirements of counterinsurgency warfare, and include the PRAL, the naval commandos, the Cazador Battalions, and the Airborne Battalion.

The formation of PRAL (a Spanish acronym meaning essentially "long-range reconnaissance patrol") units was one of the first steps El Salvador took to address the well-documented need for small unit reconnaissance patroling. The original candidates were trained in Panama in 1982 by the U.S. 3rd Battalion, 7th Special Forces Group to perform much the same mission as the U.S. Ranger companies in Vietnam: locate enemy formations for a parent unit, who would then react to the sitings. The PRAL's mission has remained largely unchanged, though they now also engage in combat patroling as well as reconnaissance. The major organizational change from their formative years is that where the initial PRAL unit was linked with the airborne battalion, today there are several PRAL companies available to meet national reconnaissance needs or those developed by brigade commanders in their areas of responsibility.

The naval commandos were initially trained by a U.S. Navy SEAL MTT and were later trained by U.S. Army Special Forces in Panama to conduct small unit ambushes and in-shore reconnaissance around La Union down to the Bay of Jiquilisco. The navy's patrol craft, particularly the Pirhana fast patrol boats, provide insertion, extraction, and fire support assistance. They represent the second major effort to produce a small unit reconnaissance patroling capability, targeted in this instance against amphibious movement of guerrillas and supplies from Nicaragua.

The Cazador Battalions were designed by a Venezuelan assistance mission based on their own experience. The organization was built around the patrol rather than the squad and included organic CA and PSYOPS capabilities. What evolved in practice were several small (approximately 300 men) battalions that were used as large reconnaissance in-force units. In that role, they lacked the support weapons for sustainability but at the same time were too large to deploy clandestinely. The late Colonel Monterrosa, one of the most successful Salvadoran commanders, combined the Cazador Battalions in his department into 700-man BIATs (Antiterrorist Infantry Battalion), which were large enough to sustain themselves for extended periods in the field while supporting small satellite unit patrol operations.

The Airborne Battalion is part of the air force and is assigned a quick-reaction strike role. It normally deploys in company strength to carry out raids or occupy blocking positions as part of a multiunit operation. It is primarily a helicopter- or truck-borne force; El Salvador has neither the aircraft, the parachutes, nor the requirement to drop more than a lean company at any one time. In most other respects, the Airborne Battalion's training and equipment is the same as any other infantry battalion's.

The air force has a role as well. Although the entire air force (like the army) is in one form or another involved in the counterinsurgency campaign, the principal fire support platforms are those designed by the United States specifically for such operations. Their use is, at least in theory, governed by counterinsurgency campaign support doctrine emphasized by U.S. Air Force air special operations advisers. These include nine A-37 Dragonfly close air support jets and a pair of AC-47 gunships ("Puff" of the early Vietnam years). There is also seven old Fouga Magisters, built originally by the French as trainers and bought for that purpose. Several have been converted to close air support platforms.

To some degree, the Salvadoran initiatives and the U.S. urgings behind them can be said to have borne fruit. More frequent small unit operations now take place under junior leaders trained to operate on their own (not a common trait in most Latin American armies), and the PRAL units and Cazador Battalions have totted up some quite productive operations.

In those units, too, an increased ability to conduct sustained, day-night operations can be noted. Strategically the government has correctly identified the major populated areas as its priority for protection, patroling, and proselytization: in insurgency, the people are the "key terrain." It has not, as some critics have suggested, left the Salvadoran-Honduras border areas of Chalatenago and Morazan entirely to the guerrillas. (Any paucity of sustained military operations in the Honduras border areas is primarily a function of Honduran sensitivity to a large-scale Salvadoran presence in those areas that is left over from the 1967 Honduras–El Salvador war.) Small unit reconnaissance, ambush, and target-destination patrols frequently help disrupt guerrilla logistical and training activities, and military operations in the guerrilla base areas support those that take place in the more populated areas. There are a good deal more of them than El Salvador is normally given credit for.

There have been reverses, of course, which many observers overuse to make critical points. The guerrillas, given the lack of government resources available to blanket the border base areas, can occasionally marshal sufficient armed strength and auxiliary support for major attacks. But a truly close analysis of those attacks will say as much for the government's resiliency and staying power as it does for guerrilla capabilities.

The raid on the Cerron Grande dam in June 1984, for example, resulted in over 100 Salvadoran soldiers (mostly local militia) killed. However, those hundred bought sufficient time against the numerically superior but poorly executed guerrilla attack that a Cazador unit was able to mount up, react, and defeat the guerrilla force before it was able to inflict any significant damage on the dam complex itself. In another case often cited as an example of the guerrillas' continuing ability to strike in force, at will, and country wide, the major training center at La Union was hit in October 1985. Many were killed on both sides, but the camp was not overrun as U.S. reporters initially said, and the next day the camp continued its training operations.

Similarly, the air force has had success in pinpoint strikes on targets acquired by ground patrols or U.S. reconnaissance aircraft. There have been isolated instances of civilian casualties, as there will be in any conflict where air power must be used, and particularly in an insurgency where a government unit under attack may well be near a populated area. But they are not the result of an indifferent attitude toward the populace by the pilots or their U.S. mentors. Again, a great deal more damage—in terms of actual guerrillas killed and, more important, disrupting and preventing large training and staging operations—is being inflicted with fewer civilian casualties than is popularly supposed.

What is popularly supposed, however, both in El Salvador and in the United States, has a great bearing on the war. That the guerrillas could carry out strong attacks (successful or not) and that civilians were killed (intentionally or not) are as important as the Western press and the FMLN propaganda organs can make them. In real terms, though, U.S. officers who have had long experience in El Salvador believe some actual attitudinal change toward the populace is taking place at the junior and mid-leadership levels, that there is an increased willingness to take the war to the enemy, and that those factors are paying off throughout El Salvador's defense forces as well as in the more specialized operations units.

This is not to suggest that all is well with the increased tactical aggressiveness of the Salvadoran armed forces, the application of it in the field, or the U.S. approach behind it. With regard to operations, there has developed a fixation on the helicopter for movement rather than shank's mare and for utilizing trails for such "footmobile" movement as is necessary. This sin is not endemic to El Salvador's or other Latin American armies either, but it does work against inculcation of the realities of sustained counterguerrilla operations.

Helicopters do have their place, of course, both in quick reaction missions and patrol drop-off. But there is some inclination to carry out airmobile assaults as the United States did in Vietnam by "telegraphing the punch" using artillery and close air support preparation of a landing zone. Here the Salvadorans' U.S. mentors may be letting them down. Some of the more successful airmobile assault operations in Vietnam were those that went in with no preparation or with fires employed as a diversion on another landing zone.

Another valid critique is that much of those successful tactical adaptations that have been made are a result of personalized rather than institutionalized action. A battalion's ability to move quickly through the bush in well-controlled but separate elements, rather than one long struggling column, for example, or a company's ability to act as a moving rather than stationary base for reconnaissance patrols and then react effectively to discoveries, depends greatly on the attitude, training, and courage of individual Salvadoran leaders. They also depend on those leaders' receptivity to U.S. operations, plans, and training advisers' counsel, and in turn, the advisers' own backgrounds and attitudes. It is true that there are indications of an attitudinal shift in junior leaders, particularly in the units considered here. But it is also true that El Salvador still lacks a professional noncommissioned officers corps, leaving officers all too often in a no-win dilemma: either do the sergeant's business themselves, which mires them in detail and limits their effectiveness as planners, mission commanders, and representatives of the government to the people, or leave it to noncommissioned officers, who may not have the necessary competence. This shortcoming is recognized by the U.S. training mission, but short-term "NCO training academy" measures cannot readily alter the fact that there is limited enthusiasm on the part of the peasantry to make soldiering a career or that the education levels for those who are so inclined is generally not at a level to make them truly effective.

It is also true that although the various in-country and out-of-country training activities run by the United States have shown results and can be directly credited with the dropping of major guerrilla tactical successes, the attitudes, backgrounds, and training of U.S. personnel have not always been those best suited for dealing with the political-military nuances of an insurgency. Some of the same false prophets the United States worshipped in Vietnam—guerrilla body counts, quantification in percentages of levels of government control, an overreliance on hardware and firepower—have been in evidence in El Salvador, and perhaps not unexpectedly. Career officers are trained to design ways to defeat an enemy force, not use military operations to serve political and psychological ends. It is antithetical to career-minded conventional military officers to think and act in terms that not only will not directly lead to total victory in military terms but relegate such a concept to a much lower priority than denying real victory, political and psychological, to the insurgents.

When the problem for U.S. operations, plans, and training advisers and the rest of the SOF trainers is seen in that light, in fact, the surprise is not that some of them may not be sufficiently attuned to the real issues at stake and the strategy required to deal with them but that most of them are. Given the limits to which pressure on the Salvadorans to adapt can be brought to bear, the subtlety with which it must be applied, and the "total victory" mindset inculcated by much of the U.S. service school system, only the credibility and countervailing influence of special operations background and training and sheer professional dedication can explain such effectiveness as U.S. trainers and advisers can claim.

For the real measure of effectiveness in a counterinsurgency is not body count. In El Salvador, it is measured by such things as the turnout for the 1984 presidential and 1985 General Assembly elections, despite guerrilla intimidation; by internal and international perceptions of government legitimacy and durability, which continue to increase; and, most of all, by the physical and psychological denial of the populace to the guerrillas. Largely through a combination of protecting population centers, targeting for air strikes and raids in less populated areas, and political reforms, that denial is developing.

Certainly there are holes. Several have been mentioned; several more could be (particularly about unit-level acquisition and processing of local intelligence indicators, something in which neither the Salvadorans nor their U.S. mentors are particularly strong). But a government facing an insurgency ultimately wins by not losing, by gaining the time to implement programs that reduce the insurgents' appeal, and by training forces to fight more effectively. El Salvador is using its time wisely. Victory may be an elusive goal in the absolute sense, but due in no small part to its own forces' contributions and those of the U.S. advisers and trainers, the government is not losing, and the guerrillas are not winning.

Other Activities

There are other showcases for U.S. application of elements of its low-intensity conflict doctrine, and there could be more in the future.

In Costa Rica, admission by the government that the fighting on its borders between Sandinistas and contras could have consequences beyond its national police's capabilities have resulted in increased U.S. weaponry aid and a small Special Forces MTT. In Panama, though the "School of the Americas" (which trained generations of Latin American soldiers) has departed for Fort Benning, Georgia, the Special Forces battalion there (in addition to providing cadres for training missions elsewhere) routinely conducts courses for Panamanian and other Latin American forces in small unit counterinsurgency-related operations.

Apart from the direct assistance counterinsurgency missions, there are others that SOF either are performing or could perform should circumstances so require. One such mission area is counterterrorism. The most obvious venue here is reaction, and both the U.S. Southern Command and the United States have SOF counterterror reaction assets available. Less obvious but more useful in the long run are the proactive measures in which SOF can assist. SOF assets have been used to good effect in such activities as critical facility black-hat penetration surveys, development of reentry and relief support plans, and training for Latin American antiterror forces.

Another element of SOF utility often overlooked by U.S. officers charged with training conventional units for operations in nonconventional circumstances is in utilizing SOF troops familiar with Latin America for the kind of area familiarization that formal area studies cannot match. SOF can look at the conventional force from the adversary perspective, provide grass-roots advice in what to expect, train in adversary tactics and weaponry, and serve as credible opposition forces in exercises. SOF in fact perform in all these roles in Latin America, particularly during the rotation of conventional units for exercises in Honduras. They can also indicate to commanders what vulnerabilities their C^3I structures may have. A former commander of the U.S. infantry brigade in Panama, for example, asked local Special Forces elements to attempt to penetrate his permanent operations center during an exercise. When they did, he had them explain in detail how they did so to his security officers.

Classic special operations missions are those in denied areas. President Reagan's stated intention to utilize SOF to train the contras addresses but one of the potential missions in relation to Nicaragua that might be assigned to SOF were the president to so direct. They include most of the publicly acknowledged SOF mission spectrum: long-range reconnaissance and selective interdiction operations, rescue operations for stranded or detained Americans, amphibious reconnaissance and raids, and PSYOPS. Air force AC-130 gunships are routinely employed in Latin America and Caribbean exercises and would doubtless find employment over Nicaragua once the air defense systems were neutralized. Certainly none of the above missions is currently authorized or being undertaken, but the continuing nature of the other SOF activities in Central America would greatly facilitate SOF transition into such missions.

As a general, final statement on the U.S. approach to low-intensity conflict in Latin America, and the U.S. SOF role in that approach, the United States and its security assistance operating arm in Latin America, the Southern Command, make maximum utilization of SOF capabilities throughout the region because security assistance, versus direct intervention, is both the peacetime and wartime game plan.

Great Britain in Northern Ireland

> A few hundred scrawls o'chaps with a couple o'guns and Rosary beads, again' a hundred thousand trained men, with horse, foot, an' artillery . . . an' he wants us to fight fair!
>
> —Sean O'Casey, "The Plow and the Stars"

> If the terrorists think they can outlast and defeat the forces of law and order then they had better think again. The armed forces, in support of the Royal Ulster Constabulary, shall continue relentlessly to hunt down those responsible for violence until terrorism is finally stamped out.
>
> —Lieutenant General Sir Robert Richardson,
> (former) general officer commanding, Northern Ireland

South Armagh is a lush, beautiful borderland between the Republic of Ireland and the rest of County Armagh in British Northern Ireland. The farmlands and towns, the hills and valleys, all commanded by a group of mountains to the north, fairly roll before the eye. The border itself meanders to and fro, and the back roads crossing it often bear no indicators that a political boundary is present. To the observer, it is a magnificent tableau of pastoral harmony. It is also an illusion.

For South Armagh is "bandit country," in all senses of the word. The interlacing of small roads with the zigzag border has long facilitated smuggling operations, and now the gunmen of the IRA and the Irish National Liberation Army (INLA) use the old infiltration routes and techniques. Freshly painted pro-IRA slogans and anti-British epithets mar walls and traffic signs, damage from recent bombings and burnings is a fact of life, and there is a sullen mood in the towns. South Armagh is a heavily Catholic, overtly anti-British section of a largely Protestant, pro-British country. To an observer familiar with conditions a dozen or so years ago, little there appears to have changed.

Indeed, in South Armagh, and some other locations like West Belfast and western Londonderry, appearances are facts. Support for the IRA is strongest, and casualties to security forces on patrol the greatest, in these areas. But one thing has changed: incidents of violence are far less numerous (though occasionally spectacular) than in the past, and in the rest of Northern Ireland, they are increasingly rare. The images of barricaded, war-torn streets with empty buildings and an embattled, despairing population on the verge of economic collapse may once have captured reality for Northern Ireland but no longer.

Accustomed to the routine of security operations, most people ignore them. They are getting on with the essential processes of living in a modern industrial state, and while bustling crowds and busy shops cannot hide all economic woes, Northern Ireland certainly has no more than the rest of the United Kingdom and fewer than those of the Republic to its south.

Such progress since the most violent times of the early 1970s represents the most meaningful kind of victory for the British in their long campaign aginst terrorism in Northern Ireland. Certainly the British have not won in the insistent, one-dimensional American sense of the word. The terrorists are still there, can still deliver highly lethal attacks, and still receive support from U.S. sympathizers, other terrorists, Libya, and the Soviet Union. The grievances of the Catholic minority that provided the original basis for IRA militancy are still there, and the enmity created by years of sectarian bloodletting is unlikely to disappear in this century. But the British have reduced the level of threat to public order to one that has given the community at large sufficient self-confidence to adapt, regenerate, and at least try to prosper. That may not yet equate to winning, but it is a success of major proportions.

At a time when many of the world's armies, particularly the U.S., French, and Soviet, are reorganizing some of their formations and resourcing priorities to build up their counterinsurgency, intervention, and other low-intensity conflict capabilities, Britain continues to wage what has become the longest counterterror–internal security campaign of the modern era. It is a campaign fraught with complexities and conflicts of emotion and burdened by the certainty that peace in the absolute sense is probably unattainable. But it is also a campaign that is yielding positive results.

Spectacularly bloody strikes by the IRA such as the March 1, 1985, mortaring of the Royal Ulster Constabulary barracks in Newry or the bombing of the veterans' gathering at Inniskilling in 1987 do not paralyze the Protestant community nor do they galvanize the Catholics into mass risings. Neither (perhaps most disappointing of all to the IRA) do they engender among the British electorate mass appeals either to overreact and return the bad old days of Bloody Sunday (Londonderry, 1972) or to capitulate and withdraw. Had not the British operations over the course of the past eighteen years succeeded in penetrating IRA inner circles and lines of communications, the IRA would be able to undertake a much wider scale of activities than is currently the case. It would also enjoy much wider internal support than it does among the Catholic community of the North and the population at large in the South.

It will be useful to analyze British operations in this sustained low-intensity environment. First, however, the environment itself must be established, and a thankless task that is. The setting for the Northern Ireland of today is born of equal parts ancient Irish history and myth, Middle Ages religious war hatreds, the romance and realities of the early twentieth century struggle for Irish independence, divisions that led to and were not closed by the Irish Civil War, and the strange bedfellows involvement of the international terrorist community, the KGB, and those Irish-Americans who, pumped up on old rebel songs and new rebel rhetoric, see themselves as part of a noble cause.

The Cause: Romance and Substance

The IRA of today—or rather, its several branches and offshoots, each of which claims to speak for the true spirit of Ireland—grew up in the wake of the 1916 Easter Rebellion in Dublin to carry on the struggle for independence. The IRA was the armed side of the Sinn Fein ("We Ourselves") party, and together they espoused the notion that all of Ireland ought to be wrested from the British.

Other parties thought differently, recognizing that since the early seventeenth century, when William of Orange's Protestant army crushed the forces of the Catholic landowners at the Battle of the Boyne, most of the ancient province of Ulster was staunchly pro-British, anti-independence, and fiercely Protestant. Believing that pushing for the northern six counties would prove a bad bargain in the long run and delay independence for the counties that did want it, these parties were willing to accept partition. The British negotiated with them, and the current boundaries were accepted by all participants. However, neither the IRA nor Sinn Fein was involved in the proceedings. They declared themselves not bound by the results, and charged that the Irish people had been betrayed. During the ensuing Irish Civil War, the Sinn Fein–IRA side found itself unsupported by the bulk of the Irish people and unable to contend with the nascent army of the Irish Free State. IRA outbreaks nonetheless occasionally occurred over the following years, North and South, until the IRA's open support of (and support by) Nazi Germany during World War II caused what little internal support remained to die out, and the "old IRA" with it.

A good many British observers today contend that their forebears helped author the "Troubles" by giving insufficient thought to the drawing of boundaries during the independence negotiations. British insistence on inclusion of heavily Catholic and historically restive West Derry and South Armagh, regions for which the Protestant majority of Northern Ireland had (and has) little use, helped ensure development of a minority that would feel itself disenfranchised and alienated, on borderlands impossible to seal off from the Irish Republic. Protestant religious arrogance and economic preeminence led over the years to social, political, and labor discrimination—conditions that naturally nurtured the rebirth of the IRA.

Legitimate Catholic grievances and pent-up frustrations would probably have found outlet in mass demonstrations anyway in the worldwide student-worker restiveness of the late 1960s; the IRA merely catalyzed and exploited them. Violent Protestant mob attacks on Catholic demonstrators provided exactly the atmosphere and recruiting impetus the IRA had dreamed of, and it declared it would protect the people. By the end of 1969, its support among the Catholic community was almost universal. But the IRA at that time had competition, which also enjoyed a measure of Catholic support, in the form of the British Army. The Royal Ulster Constabulary was insufficiently armed, staffed, or inclined to halt Protestant attacks on Catholic demonstrators, so the British government deployed regular army battalions to protect the Catholics and separate the two warring communities. This origin of British military involvement in Northern Ireland is usually forgotten by most adherents of "The Cause," particularly those in North America.

The IRA, faced with this threat to its status as protector, set out to turn the competition into the people's adversary by engineering increasingly violent confrontations with British troops by small groups of thugs and children. They were also again assisted indirectly by the Protestants. When the British Army intervened, the Protestant mobs retreated and did not engage in confrontations with British troops. Thus, when the British responded to attacks, as they had to, it looked as the IRA hoped it would: British troops supporting the (Protestant) establishment against the Catholic minority struggling to be free.

It was masterfully, if murderously, done. It succeeded in bringing about virtual military occupation conditions, near paralysis of economic life in many areas, and occasional overreaction from sorely beset British Army units, as well as resumption of sporadic attacks by Protestant extremists. It succeeded in legitimizing IRA rhetoric and existence in many previously undecided hearts, in both Ulster and the Republic. It also succeeded in ensuring the demise of any hopes for peaceful union by exacerbating and solidifying ancient prejudices. As the wife of a Republican interned in Long Kesh prison remarked when asked by a British member of Parliament (who was himself at that time a part of a growing "troops out" movement) what she thought would happen if the army left: "We'd all be murdered."

The question of why the IRA, or its offshoots, has strived so mightily to ensure a high probability of a general bloodbath if the British Army did pull out is complex. First, by having engineered such conditions, it has virtually prevented its symbolic adversary—the British Army—from leaving, thereby ensuring its own continued existence (though most of the IRA's victims are not British troops but off-duty policemen, militiamen, and innocent civilians). Second, its own post-British visions (to the extent they are formed at all) involve a united Marxist-socialist state markedly different from the current governments of the North or the Republic; such a state could only be born out of a major, violent upheaval, which would bring down the current political system of both North and South. A third consideration is that there is no single IRA entity any more. The "official" IRA early on was Marxist oriented but came to abhor the reversion to classic terrorism relied upon by the Provisional wing. The Provisional wing, or Provos, have in turn been insufficiently violent in recent years for some members, who have broken off to form the openly leftist INLA, which espouses a doctrine of anarchy and worldwide revolution indistinguishable from that of other European terrorist organizations. None of these groups could expect support from sympathizers in the United States or their other suppliers if their revolutionary enthusiasm were seen to be flagging.

There is, however, another rationale for the continuance of violence. Its most generic, quintessentially terrorist side is best expressed by the old Irgun slogan: "I fight in Palestine, therefore I am." But its less formed, uniquely Irish side has been called by Conor Cruise O'Brien, the distinguished Irish playwright, historian, and diplomat, the "romance of failure." In effect, it is the association with the legends and traditions of the long years of struggle, not the results of the struggle itself, that provide satisfaction for the Provos (and their supporters) of today. It is the striving, not the doing, and the martyrs, not the makers, that hold meaning. The view stated by O'Casey's hero in the passage quoted at the beginning of this section is more than the ennoblement of terrorist tactics; it encapsulizes the self-image of many of "The Cause's" fighters and supporters. Ultimate victory—which, if one believes the rhetoric, means peaceful union of North and South—would end the striving and provide no new access to the legends. To many hearts, that possibility is more hateful than anything. The romance can continue only with blood shed in the course of struggling.

Terrorist Operations

A principal tactic in the early years of the Troubles was to generate some mob rock- or Molotov-cocktail-throwing actions against British patrols, which would bring backup forces. Gunmen would then shoot at the British from safe roof or window vantage points, provoking retaliation and civilian casualties. Bombs were planted by sympathizers in restrooms of bars and clubs frequented by Protestant militia personnel or off-duty policemen and soldiers. In the wake of Bloody Sunday, in which British troops shot thirteen unarmed demonstrators, and Operation Motorman, during which British forces cleared out Republican "no go" areas in Londonderry and Belfast, squad-sized ambushes were not infrequent. Potential support in the form of safe houses, temporary hiding places, caches, and transportation was well organized throughout Northern Ireland's Catholic communities, and gunmen could for a time safely gather, plan, and conduct an operation and then virtually disappear. In the early days, before Soviet, international terrorist, Libyan, and North American sympathizer support began to introduce modern weaponry and demolitions, IRA armament was largely the stuff of the old rebel songs and stories: British Webley .38 service revolvers, American Colt .45 automatics and Thompson submachine guns hoarded over the years, and the odd Enfield, MAS-36, or Mauser rifle. Demolitions were gelignite, guncotton, and some commercial dynamite, along with World War II–vintage American hand grenades. The emotionalism that swelled IRA ranks brought in mixed skill levels, and the IRA provided little training. Thus, marksmanship was not particularly noteworthy, and demolitions caused almost as many casualties among the erstwhile bombers as they did among their intended targets.

As the years have gone by, a number of changes have taken place. The old cast-off weapons have been replaced by U.S. M-16s or the civilian version of the same weapon, the AR-15; .357 and .44 Magnum revolvers; Beretta-12 submachine guns; AK-47s; and reasonably reliable Soviet detonators. Manually planted time bombs have been succeeded by radio- and command-wire-detonated charges of tremendous power. Ambushes in urban areas still take place, but more often gunmen living in the Irish Republic come across the border in South Armagh or Derry, take a few shots, then race to a waiting car and speed back to safety. As operations in urban areas have become increasingly risky due to British intelligence penetration of sympathizer groups, numbers of local activists have dropped off. This has left a hard core of adherents, many of whom have been trained in PLO camps and who have served (and survived) for years. Thus, while the scope and quantity of operations have declined, the sophistication and spectacle of the attacks have increased. Moreover, while once the principal targets were British troops, they are often now soft targets such as Royal Ulster Constabulary and Ulster Defense Regiment reservists and judges. As operations against even these targets in Ulster have become more difficult to mount with fewer in-country support assets, the IRA has turned to targets outside Northern Ireland. The attempted assassination of Prime Minister Margaret Thatcher, the bombing of British officers' clubs in England, the Household Cavalry, the commander of the Royal marines, and Harrod's department store testify to the IRA's continued ability to strike. The INLA has also proved itself capable to killing defenseless people in the bombing of Lord Mountbatten's boat and the slaughter of the Anglican church congregation in Ballykenny.

Mass ambushes on security forces no longer being feasible, the IRA has turned to planting command-wire bombs weeks in advance of the desired demolition date, hiding the wire, and then returning to make a quick hookup when intelligence advised of a particularly lucrative target. These bombs are placed not only in the roadbed but in the walls of stone fences or buildings abutting the roads. The effect is to create not only explosions of tremendous force, which can concuss those inside an armored Land Rover or truck (if not incinerate it), but also a massive fragment-scattering effect from a stone structure at head level. Two of the major losses to British forces in recent years, one in a Scots Guards patrol and another in a patrol from the Parachute Regiment, occurred in this way.

Possibly the most ingenious device developed by the IRA has been the truck-mounted multiple mortar. This is a collection of cut-off propane or oxygen tanks, bracketed together on the back of a flatbed truck, braced against the truck's cab at the desired angle, then rigged with homemade propellant charges and loaded with armed mortar bombs. These trucks are parked in areas known to be relatively free of British patrols but fairly close to a fixed installation and a command wire run out a mile or two. A covering tarpaulin is left over the tubes to prevent aerial detection. At the appointed time, a helper (either brave or stupid) may pull off the tarpaulin and then beat a hasty retreat. The gunner, neither brave nor stupid, sets off the charges either all at once or individually from a safe distance. For all its apparent crudity, the device is reasonably accurate, and its rate of malfunction (and dead helpers) is remarkably low. It has been around for several years (despite its recent discovery by the American media); the 1985 attack on the Newry Royal Ulster Constabulary station is only its most spectacular and successful instance of use. In late 1984, in fact, the British fort in Crossmaglen, South Armagh, was struck by six-round barrages twice, both of which caused structural damage but no deaths or major injuries.

British authorities estimate current Provo active terrorist strength at 400 to 500, or less than half of their mid-1970s number. Active support (in Northern Ireland, the Irish Republic, and overseas) is reckoned around ten times that, while another factor of ten, or 50,000, is estimated as a passive support figure. The INLA is thought to number fewer than a hundred actual activists, with much less proportional popular support than the IRA but with good contacts with Palestinian and European terrorist organizations.

The IRA Provos receive money and arms direct from the United States, though as a result of the North American Arms Control Act of 1968 and the relative ease of getting arms elsewhere, most support from the United States is in the form of funding. Much of it comes, U.S., British, and Irish government officials believe, from the Irish Northern Aid Society (NORAID). NORAID claims (and may actually believe) that its fund raising goes to aid homeless children, but evidence to the contrary is sufficiently convincing to discount most NORAID protestations. Health and homeless-children care are guaranteed and virtually free in Northern Ireland, as in the rest of the United Kingdom, and too many arms shipments have been traced to NORAID funds.

While some arms shipments have been bought from international arms dealers in the West, the laws of most countries require that purchasers of case lots of military arms be representatives of other governments. Thus, such purchases have been illegal, though they may have been covered by forged end user certifications (available from corrupt diplomats) ostensibly furnished by one country or another that the purchasing agent is indeed a representative of that country. Less troublesome, though still risky, is getting arms through terrorist sources or directly from Communist suppliers. Shipments from Czechoslovakia and the Middle East have been discovered and seized, though many more get in through bribed customs officials in Ireland, France, Portugal, Holland, and Belgium. High seas direct deliveries are increasingly rare and are risked only when needs are so urgent that the time-consuming multinational smuggling process will not do. It was such a shipment from the United States that the Irish Navy seized in 1984.

The most significant support linkage in recent years has been the one established with Libya. Libyan support originally surged in the early years of Qaddafi's rule and then waned for a time as the Libyan leader found other enthusiasms. In the wake of Libya's expulsion from Britain after the killing of a policewoman by Libyan People's Bureau diplomats and Britain's support for the 1986 U.S. bombing attacks, Qaddafi apparently decided to renew his support for the Provos. In October 1987, a ship was seized by the French carrying a large quantity of Soviet-manufactured assault rifles, mortars, machine guns, rocket-propelled grenades, and most ominous, hand-held surface-to-air missiles. This shipment was described by the seized skipper as the fifth of six. A massive Irish Defence Forces sweep turned up little, causing many in the United States to doubt the existence of previous shipments. But the British and Irish governments did not discount the story, nor could they. Even one shipment like that found on the *Eksund* would exponentially raise the qualitative and quantitative striking power of the Provos. Their concerns were apparently borne out in January 1988 when a cache of AK-47s, rocket-propelled grenades, and RPD light machine guns was discovered by the Irish police in County Donegal.

The INLA is thought to receive no aid from North America at all. Without funding support and arms channeled to them from European and Middle East terrorist groups, it would likely disappear. In fact, as a result of internal friction and differences with the IRA, it has become an almost irrelevant factor.

External support aside, the terrorists have adapted to reduced operational scope and public support by conducting fewer but better executed operations, with better equipment and more experienced personnel. But what led to the necessity for the adaptation? Why can they no longer operate at will, with people living in the same area in which attacks are carried out? Why must attacks be launched from and return to the Irish Republic or take place overseas? The effectiveness of the British response is primarily responsible.

The British Response: From Motorman to the "The Way Ahead"

As the demonstrations, barricade building, and mob violence that began in 1969 quickly escalated beyond the Royal Ulster Constabulary's capabilities, the military general officer commanding was given control of all security operations. Principal equipment on troop patrols depended on commanders' and intelligence officers' estimates. If riots or demonstrations were expected, platoons deployed carrying riot shields, batons, and gas guns, with standard weapons slung (except for designated counter-sniper teams). At other times, Land Rovers or trucks following platoon box patrols carried the riot equipment, while the troops themselves patroled as if in a war zone. Rules of engagement were strict, and instances in which they may not have been adhered to were in fact exceptions, not the rule. Any violations at all, however, became leading news stories and grist for IRA propaganda mills, and bloody crowd reactions could be expected.

On August 9, 1971, the British carried out a fishnet arrest and internment operation, detaining and interrogating virtually all persennel on barricades, found rioting, or otherwise known (or suspected) to have Republican sympathies. The techniques utilized came under some criticism from human rights organizations, and it may well be the British created more Republicans than they locked up. It is also true, however, that a good deal of valuable information on rebel leadership, transportation routes, and arms caches was developed. At this time, there were about 14,000 British troops in Northern Ireland.

Almost a year later, July 31, 1972, the levels of British troop strength and IRA control of Catholic neighborhoods had reached what would be their respective peaks. The IRA had established "No Go" areas, which meant Catholic neighborhoods of Belfast and

Londonderry, as well as areas of rural South Armagh, were off limits to Protestants, the police, and British troops. In effect, it was a dare. The British could either accept what amounted to rejection of Crown authority and unilateral declaration of Catholic independence or move to end the challenge in such a massive way that civilian casualties were bound to ensue.

There was no choice for the British, of course, and Operation Motorman thus commenced. Troop strength in the province had by now increased to 21,000. The British moved against the barricades, broke them up, and when fired upon returned the fire. Gas and riot batons were used extensively; the streets were cleared and held virtually as occupied enemy objectives. There was no escaping the visual image Motorman created of brute force being applied against largely unarmed civilians, no matter how extreme the provocations by these civilians were and no matter how comparatively humane the response. Casualties did occur; arrests were made.

In the long run, some results became apparent. Given that the British had no choice under the circumstances and that the IRA would make the most of the event, the British also made the most of it. The IRA had been penetrated on several levels, not extensively but enough to provide some direction to Motorman that the IRA had perhaps not counted upon. Moreover, it is clear from propaganda printed before Motorman for distribution in its wake that the IRA anticipated a much more heavy-handed and bloody response than was in fact the case. Blood was shed and animosities exacerbated but on nothing like the scale the IRA apparently expected.

In urging the "No Go" zones, the IRA expected to take some losses, not just among the Catholic communities but in their own ranks. As it turned out, the British response caused fewer general civilian casualties and netted more IRA activists than the IRA could have anticipated. For the IRA, the results of Motorman did not justify the degree to which they had placed themselves at risk. It may be they actually believed the time and prospects for a general rising of Catholics were right, though again, the preprinted propaganda makes clear they expected it to be crushed. More probably, the IRA expected a British response that would earn it the condemnation of the world at a low cost to itself. It was disappointed on both counts.

On balance, the "Day of Internment" and Motorman proved useful to the British, for they provided opportunities—indeed, necessities—for application of massive force in quelling what had become insurrectionary conditions. In doing so, an infrastructure untouchable through normal legal procedures or in simply cordoning off troublesome areas became vulnerable. While the British certainly did not, either on the "Day of Internment" or Motorman, break the back of the IRA, they did begin to penetrate it as a result of the former and fragment it during the latter. By literally searching every Catholic dwelling and building during Motorman, the British forced the IRA to flee; by staying, the British broke the stranglehold the IRA had previously on daily living in those areas. The British were no more loved; they were, however, more in control.

After that, the British maintained the pressure, and the IRA was unable to regain either its former level of control or freedom of movement. Statistics do not tell the whole story, but they can show some results. In 1972, 17 policemen, 129 soldiers, and 321 civilians died in Northern Ireland. The following year, the figures were 13 police, 66 troops, and 171 civilians, and the year after that 15, 25, and 166, respectively. The downward trend has largely continued since then, with some blips reflecting spectacular multiple-casualty bombings and the deaths associated with the hunger strikes in 1982. In fact, the incidents of violence had dropped off so sharply in the wake of Motorman that by 1975 the British had withdrawn a third of the troop strength, and in 1977 the Royal Ulster Constabulary was restored to primacy in carrying out security responsibilities. This occurred as part of a program called "The Way Ahead," in which the army is considered the junior partner to the constabulary.

Relative success during Motorman does not in itself account for the dropping off of IRA control and operational scope. Rather, the consistent (and persistent) application of proactive and reactive measures at the tactical and policy levels, reduction of military visibility as quickly as conditions allowed, and conduct of aggressive, combined police-military intelligence operations have together imposed severe limitations on IRA operational flexibility and brought about a return to normalcy in most of the province. These elements bear further consideration.

Multidimensional Intelligence Operations. Combined police-military intelligence activity is a feature of all British counterinsurgency operations, though since this one is on British soil, there have been some departures from the overseas pattern. The Royal Ulster Constabulary's Criminal Intelligence Division and Special Branch, the same divisions of the main police force in England, the national counterintelligence force, and the national foreign intelligence service, all have roles. Crossborder and in-border tactical intelligence is also collected by the SAS, unit patrols, and electronic warfare detachments. Nationally derived intelligence is pooled in London (which has its own problems with the IRA) and then provided to the constabulary's Special Branch and to Headquarters, Northern Ireland's intelligence section. At these locations this input is synthesized with information available from police and military collection means, shared between the military and the constabulary as appropriate, and acted upon. Prior to 1977, the military ran the joint effort; now it is a cooperative one, as are all other aspects of constabulary-military operations.

Civil versus Military Authority. It is British preference to place essentially police powers in the hands of the military for only the minimum time required. While the Royal Ulster Constabulary are no more loved by Catholics than the army (over 300 of them have been killed since 1969), they are representatives of the civil power rather than a military occupation. By restoring that civil power, the British deprived the IRA of a powerful propaganda element. At the same time, they have maintained the military umbrella under which civil power could function. Many other nations, faced with urban insurgencies, have encountered the dilemma of providing for sufficient emergency powers to allow security forces to deal with the insurgency without compromising essential freedoms. The British resorted to emergency powers but took care to see that they were already sanctioned by law or countenanced by Parliament. Those powers have not been abandoned, but they are enforced by civil authorities with military backup, not the other way around.

Cooperation with the Republic of Ireland. It has helped the British effort that the government of the Republic of Ireland also considers the IRA terrorists and outlaws. The Republic has had its own problems with terrorist incidents on its soil, and the government has been denounced as traitors by the IRA.

The Republic carries out some military border patrols, and the national police, the Garda Siochana, has responded when advised by the Royal Ulster Constabulary of suspects fleeing across the border. The Republic has even found, seized, and extradited high IRA officials in the past several years. Officially relations are said to be excellent.

At the working level, that appears to have largely been the case, though there have been chilly periods. As a result of a botched extradition request and some insensitivity by Prime Minister Thatcher to New Ireland Forum suggestions in 1984, there was a noticeable cooling off of relations between the Garda and the Royal Ulster Constabulary. Another rift grew out of the British judicial decision in January 1988, absolving constabulary "shoot-to-kill" suspects, which ran counter to Irish popular opinion. There is also some animosity over the surfacing of British intelligence operations in the Republic in London newspapers.

It must be remembered that while the Irish government condemns the IRA and not many Irishmen actually support it, neither are Irishmen by any means pro-British. Any government cooperation with the British must be handled delicately and in accordance with the Irish laws. When it appears to the Irish that the British have forgotten these points, cooperation drops off. Nevertheless, the fact remains that whereas once IRA offices operated openly in border towns like Dundalk and Drogheda, the IRA must now operate nearly as furtively as in Ulster. For the British part, whereas British operatives in the Republic may once have had a more flexible brief in dealing with known IRA personages, they now can do no more than observe, report, and hope that the Irish government will take action.

Adapting to Circumstances: Manpower and Organization. The British have devoted considerable thought and effort to ensuring that the composition and missions assigned to various components of the security forces were appropriate to the times and to the prevailing conditions of the areas patrolled. The Royal Ulster Constabulary is trained and equipped to carry out patrols in low- to mid-threat areas on its own and with army support in high-threat areas.

Energetic attempts are being made to recruit Catholics for the constabulary, though with little success; the IRA has sworn to kill Catholic policemen as "traitors" and occasionally does. Consequently, Catholic membership on the 8,000-strong constabulary is less than 2 percent. Over 40 percent of the overall military strength consists of the 7,000-strong Ulster Defense Regiment (UDR), a unique force in the British Army. The UDR is commanded and staffed by regular British officers at the upper levels but manned primarily by locally recruited personnel. It is about half full time and half part time, is never deployed outside Ulster, is never used in riot control duties, and does not patrol the high-threat areas of West Belfast or South Armagh. Six regular British battalions serve two-year tours at various locations around the province, while two battalions—called the "Roulement" units—serve four-and-a-half-month tours in the high-threat areas. The Ulster Defense Regiment and two-year-tour battalions carry out primarily vehicular patrols to show the flag but come into direct contact with the civil populace only in support of constabulary operations. In South Armagh and West Belfast, the worst areas, elite regular battalions still operate as the senior partner in the army-constabulary relationship. Constabulary personnel go out in army patrols to provide arrest authority or carry out other police activities but live at all other times inside Army–Constabulary stockades.

The British Army does recruit other regiments from Northern Ireland—the Irish Guards, the Royal Irish Rangers, the Queen's Royal Irish Hussars, and the 5th Royal Iniskilling Dragoon Guards— but these do not normally serve in Northern Ireland. Similarly, while there are some 4,000 territorials (reservists) resident in Northern Ireland, they do not serve with security forces.

Adapting to Circumstances: Tactics and Equipment. Military foot patrols have become increasingly rare, except in West Belfast and South Armagh. Where they do occur in built-up areas, they are deployed in four or five four-man "bricks," usually commanded by a corporal. Each member has radio communications. These bricks satellite off the platoon headquarters patrol, which would typically have a Royal Ulster Constabulary contingent with a dog. In the countryside, bricks operate farther apart and are normally dropped off by helicopter. Vehicular patrols are considered appropriate for most urban situations since they cover ground faster, are not as conspicuous in urban crowds as foot soldiers in battle dress, and the vehicles provide protection against most threats. In South Armagh, vehicular patrols are no longer mounted since virtually all roads are abutted by stone fences and building walls from which bombs have blown vehicles to bits. Patrols there deploy from the forts by helicopter and return the same way or on foot. Each fort does have its contingent of armored Rovers or Humbers to assist in crowd dispersal or conduct rescues when necessary, but their use is extremely limited.

In areas where demonstrations and riots are still possible (such as Belfast and Londonderry), a wide assortment of armored vehicles is used. Armored Land Rovers are used as command cars and patrol vehicles by the constabulary, and Saracen armored cars provide protected firepower when required. The ground workhorse of the British forces is the Humber one-ton armored truck, or "Pig," configured in several ways for several jobs. There are bomb disposal, ambulance, and signals Pigs and even a Flying Pig, which has swing-out armored panels behind which troops can advance during a riot and which can provide an effective roadblock. In South Armagh, the lifelines are big Royal Air Force Wessex and light army helicopters, which conduct all personnel transfer, patrol deployment, resupply, and other administrative travel requirements. The "Roulement" battalion's troops live and operate out of facilities surrounded by high metal walls and observational and firing blockhouses. Observation posts are located on the high mountains around South Armagh, and foot patrols daily saturate the countryside. When major damage to a fort occurs from a bombing or mortaring, the repair personnel and equipment will be deployed if at all possible by air. If not, a major armored column will be sent out, covered from the air and preceded by foot patrols to detect command-wire bombs.

Use of Deception and Surprise. Of no little value in placing the IRA on the defensive in recent years has been the success of SAS and "Roulement" battalion covert operation patrols in observing and ambushing IRA elements. Both units typically send out four-man patrols, which will go to ground near a suspected crossing point or a likely command-wire target. They will wait for days on end;

should the effort pay off, the quarry will either be ambushed or observed as they make contact with others. Also successful on a number of occasions have been instances where known IRA operators across the border were given information on easy and lucrative targets, only to find that the information had been provided by British intelligence and they themselves had fallen into an SAS ambush. Such tactics would be impossible without penetration of IRA circles in both the North and the South and intimate knowledge of IRA psychology.

Perhaps the key factor in the measured success of British operations is demonstrated commitment. In almost twenty years, the worst the IRA could do from an insurrectionary standpoint it has already done. Neither a general Catholic rising, nor panic among the British populace, nor especially condemnatory world opinion has developed. It has become obvious to all concerned that the IRA will not drive the British out. That being the case, the population of Northern Ireland has adjusted to a level of violence that, while certainly not acceptable, is at least tolerable. Protestants and Catholics still distrust and hate each other, but the security forces are a credible deterrent to most sectarian mob violence; without violence, a sense of normalcy has returned to most of the province. Whether disliked or appreciated, the troops are a fact of life, an ever-present reminder that the civil power is not operating without support and therefore cannot be held in contempt. The credibility of the civil power, thus strengthened, makes it more effective and therefore its need for troops more limited. The result is that open, violent incidents of Protestant extremism have been all but eliminated, and the IRA has been reduced to generating isolated spectacles. Far from final victory, far more from true peace—but also a long way forward from flaming streets, "No Go" zones, and the despair prevailing at the time of Motorman.

6
The Role of Special Operations Forces among the Neutrals: Unique Challenges, Unique Responses

The neutral nations of Sweden, Switzerland, Austria, and Ireland publicly proclaim a position of nonalignment in the adversary relationship between the Eastern and Western power blocs. Undoubtedly they fervently hope that this nonalignment might be carried into any general East-West conflict, and while in the case of Switzerland there may be some rationale for that hope, contingency planning and national armed force structures reflect more realistic thinking. The Swiss, the Austrian, and the Swedish defenses are geared toward assaults from the Eastern, not the Western, powers, though absent a state of hostilities, those nations will likely regard violations of their territorial integrity from any of the NATO powers with as much unfriendliness as they would an encroachment from the East. The Irish face a unique problem: external support for an internal element (the IRA) with the stated intent of supplanting the current body politic with a Marxist-oriented socialist state. Doubtless the Irish view the potential for both destabilization and external support as being even greater in an atmosphere of general East-West hostilities.

Thus, while these nations are neutral in peacetime international political life and certainly not to be thought of even in wartime scenarios as NATO adjuncts (if they are not attacked, they have no reason to fight), cultural and geographical realities have resulted in planning and, indeed, some mutual training activities, which reflect a recognition that in a general war, side choosing is likely to be forced on them and that that choice will involve conflict with the East, not the West.

In the real and more immediate sense, however, some choices have already been forced. Because of their positions as hosts of international organizations and conferences, Austria and Switzerland possess at any given time an abundance of high-visibility targets for international terrorists, and both nations have seen themselves become part of terrorism's stage. Ireland also faces an immediate problem—the IRA and its supporters, whose dogma calls for elimination of the current Irish Republic as well as British withdrawal from Northern Ireland. Sweden has faced no significant terrorist threat but recognizes the scope of terrorism's reach and menace, and therefore, like the other neutrals, it cooperates with NATO allies in exchanging information and other forms of assistance in that area. Moreover, Sweden has faced persistent Soviet attempts to reconnoiter the approaches to its sensitive naval installations.

How the neutrals have responded to the current actual and potential wartime threats in terms of special operations capabilities and missions presents as wide a spectrum of approaches as those of the NATO allies.

Sweden and Switzerland are unique in that while they have dedicated military SOF, they are almost entirely composed of reservists who rotate into and out of the shell establishments during their initial training and annual service. Sweden actually has three forces, two of them naval. One, the Coastal Ranger Company, is subordinate to the Naval Coast Artillery and is tasked with supporting harbor and strong-point defense through reconnaissance, ambush, and small attacks on enemy amphibious lodgements. The second is a force of attack divers whose training is similar to the Coastal Rangers but who also have missions involving mine clearance, countering enemy amphibious capabilities, and underwater reconnaissance and sabotage operations. The army force is a parachute Jaeger company, trained and organized like its Norwegian counterpart. The naval forces' mission spectrum is not as broad, and thus the limited training time available can be dedicated to realistic, completable tasks. Accordingly and unsurprisingly, the naval forces are rated better at their tasks than are the parachute Jaegers at their rather more complex menu of missions.

The Swiss military special operations capability is vested in one special missions company of parachute grenadiers, who have typical reconnaissance, sabotage, and resistance support missions set in a stay-behind frame of reference.

Neither nation has attempted to create strong, centralized national counterterror forces. In the Swedish case, there has been no perceived need—not because there is no perceived threat but because the special unit of the Stockholm Police organized for counterterror operations is, in effect, placed under operational control of the national Ministry of Justice for counterterror operations anywhere in the country. Thus it is regarded as a national, versus purely municipal, asset, and further forces are not viewed as necessary.

In the case of the Swiss, the concept of a national counterterror force as part of a national police agency has been the subject of recurrent debate. Both leftist fears of seeing a national antiriot capability developed along the French model and cantonal insistence on preservation of cantonal authority have been the main roadblocks to development of such national capabilities. In the meantime, local counterterror incidents are handled by SWAT units of cantonal police units, and major incidents are handled by groupings of such units from the several cantons under mutual assistance agreements.

It can be argued that the potentially uneven training received by separate cantonal police organizations militates against confidence in any cohesive, organized joint response. To minimize this recognized unevenness, the major cantonal elements meet annually for a joint training session. The German antiterrorist force, GSG-9, periodically assists in conducting this annual training. It is obvious, then, that the Swiss federal government has recognized the shortcomings of decentralized training and operations and has undertaken both operational and legislative initiatives to remedy them. That the measures that do exist are at least to some degree effective was demonstrated in the resolution of several incidents in 1983 and 1984.

Austria's military cannot be said to have developed a significant special operations capability either in the widest interpretation of the term; however, they have modeled their slim reconnaissance resources on the German Fernspaehkompanien. Indeed cooperation between the two nations in training is routine and ongoing, though it is not on a large scale and the Austrians understandably wish to keep its profile low. Like the Germans, the Austrians do not acknowledge any offensive orientation and therefore by extension do not acknowledge any cross-border special operations requirements. It is unlikely, however, that Austrian reconnaissance units are unfamiliar with basic sabotage and interdiction techniques.

In the area of counterterrorism, Austria has long perceived a threat in a connection with its status as the host of the Organization of Oil Producing and Exporting Countries (OPEC), and its operation of Jewish refugee facilities. It formed a unit out of its gendarmerie in 1973 specifically to provide protection for dignitaries and facilities, but in the wake of the 1975 mass OPEC oil minister hostage taking and the GSG-9 operation at Mogadishu in 1977, Austria concluded that improvements were necessary. Thus was formed the Gendarmerieeinsatzkommando (GEK) (Gendarmerie Special Operations Command).

GSG-9 played a major role in the initial organization, equipping, and training of the GEK; however, the GEK is not the mirror image of its German counterpart. The Germans are organized into units built around groups of a basic four-man team; the Austrian basic unit is a three-man element. The Germans are heavy on German-made equipment; the Austrians rely on Austrian (Steyr), Israeli, and Belgian weapons and equipment, though they do utilize German special-purpose vehicles similar to those used by GSG-9.

The GEK has not been used in live situations yet, but it is nonetheless highly regarded (perhaps unsurprisingly) by its German mentors. It also apparently has wide public acceptance and official backing. It has carried out a number of highly realistic, arduous exercises successfully, all of which have been extensively publicized by the government. That publicity, it is felt, confers both deterrent and public confidence benefits.

The Irish are, in many respects, in a class by themselves. Overt conventional aggression is unlikely to manifest itself in anything like short- or no-notice circumstances; for such a threat, time would likely exist for the Irish to transform themselves into a nation at arms. Much more likely are the possibilities that any Soviet destabilization or sabotage offensive launched against the United Kingdom might well include preemptive destruction (with or without increased activity by the IRA) of facilities in Ireland that the Atlantic Alliance could potentially utilize in the fight for the North Atlantic and of increased political violence in the Republic attending events in Northern Ireland. Indeed, the 1986 agreement between the United Kingdom and the Irish Republic giving the latter a greater role in Ulster's internal affairs in return for recognition of British sovereignty has inflamed extremists of both left and right and resulted in renewed threats by the IRA to the government of Ireland.

Thus the Irish must not only deal with a degree of ongoing terrorism in a delicate atmosphere of cooperation with the unpopular British, as well as plan for dealing with externally launched (or supported) subversion in a general conflict; they must also plan for large-scale political violence arising out of renewed exacerbations of dormant but by no means unsubstantial divisions over the Ulster question within the Irish population. The emotional currents involved run deep; indeed, the postindependence civil war was fought over the issue. The resurgence of the IRA in the past twenty years has captured nowhere near the level of domestic Irish popular support as is imagined overseas (particularly in the United States), and its antigovernment, Marxist rhetoric further alienates many Irishmen. But the "Brits out" passion is a banked fire that can be brought into open conflagration if the bold British-Irish experiment in the North fails.

To meet the challenges manifested by these unique circumstances, the small Irish Army has developed the Army Ranger Wing. Its actual creation was inspired in the late 1970s by the kidnapping and extortion of a Limerick industrialist by several IRA personnel. A major hostage-barricade situation ensued, in which the Garda Siochana felt military assistance was required and called upon the army in accordance with the "aid to civil power" provisions of the 1954 Defense Act. Although the incident was resolved when the perpetrators surrendered, the fact that no military unit was specially detailed, trained, and equipped to deal with such situations was not lost on either the army or the government.

Because principal responsibility for the resolution of a terrorist incident belongs to the Garda, the latter also began upgrading its dedicated counterterror responsibilities. The Army Ranger Wing's primary mission is listed, therefore, as "execution of specialist military tasks in conventional military operations," specifically long-range patrolling, raids, and support for friendly stay-behind elements. Its secondary mission is essentially to serve as a specialized training source for the rest of the army, both in terms of courses that the wing conducts in specific areas (patrolling and demolitions, for example) and the wing veterans it returns to other active (or reserve) units who carry the specialized skills with them. Its third-listed mission is "aid to the civil power"—both proactive and reactive antiterror operations.

The wing's training resources reflect the constraints of a small (and not abundantly affluent) popular tax base and deep dislike for the concept of a standing army. The entire Permanent Defense Forces number under 15,000 personnel. Thus, rather like Denmark, Ireland expects to get maximum utility for minimum expense out of its small special operations establishment. Individual training is rigorous (only one out of five candidates successfully complete it) and morale is high. However, the wing has to rig small military and Aer Lingus aircraft for airborne operations (the air force has no plane capable of carrying the entire unit), and it has not had the opportunity to test its capabilities in a major exercise postulating centralized launch and control of multiple teams in coordinated denied-area missions.

Faced with such constraints, the wing has improvised imaginatively. Like its continental brethren, it has also availed itself of the training establishments of others, though not on a large scale. In its early formative period—indeed, when the concept of a special operations capability was far from realization—several army officers attended the U.S. Army Ranger School. Since the wing's activation in 1980, selected personnel have trained with the Dutch Marines, the French GIGN, and GSG-9. These areas of training emphasis have applied primarily to the counterterror side of the wing's mission, but the individual and some team skills involved are applicable to other special operations missions as well.

The wing is equipped with standard Irish Army communications and vehicles. In the former area it is likely to find itself at a disadvantage if it is ever called upon to perform its wartime mission because tactical communications are not usually compatible with long-range reconnaissance or other special operations requirements. In weapons, the wing relies primarily on Heckler & Koch rifles (5.56 and 7.62 mm) and submachine guns and on Smith & Wesson pistols, though other brands are utilized for familiarization, testing, and special purposes. It trains in close quarters battle drill, reconnaissance techniques, small boat operations, parachute operations, and mountaineering.

In addition to augmenting the Garda's Special Task Force in raids on suspected terrorists hideouts and on border area searches, the wing also assists in dignitary protection. The 1986 European Economic Community meeting in Dublin was an example of the type of occasion where Garda capabilities may be insufficient. Those capabilities, however, are not inconsiderable. The Special Task Force is a well-organized, elite unit developed much for the same reason as the FBI's hostage rescue team in the United States: to extend the civil power's capacity for resolving serious hostage-taking incidents and thereby reducing dependence on the military.

Unlike the FBI's rescue team, however, the Special Task Force has been forced by the nature of the unique political conditions on the island into proactive paramilitary operations, including successful searches and seizures along the border and the 1980 raiding and termination of a Provo training camp in a remote area in the southwestern part of the Republic. Accordingly, the Special Task Force must—and does—take steps to ensure interoperability with its military counterpart in areas beyond reduction of hostage and barricade incidents. There are challenges to that interoperability, however; the wing and the task force utilize different communications systems, as well as different weapons. However, cross-assignment of liaison officers during field operations serves to mitigate greatly some of these challenges.

Each of the four nations has had to make decisions about dealing with the realities of geographic location, probable Soviet courses of action, internal subversion activities, or some combination thereof, that have necessitated both planning and active (albeit limited) cooperation with various NATO powers. One must beware of overstating the case; a perceived common threat regarding terrorism, and the intelligence and training exchanges resulting from that perception, should not be blown out of proportion. Still, neither the projected wartime threats nor those currently manifested within the "shades of gray" emanate from the Western powers. Cooperation is low key, low visibility, but it is there. It is appropriate that the special operations community is the medium for that cooperation.

7
The Role of Special Operations Forces in Antiterrorism and Counterterrorism

The periodic bludgeoning of the public peace so much in vogue has been attended by increasing sophistication and lethality of terrorists' means, an equally increasing diversity of ends toward which the terrorists employ those means, and, infinitely more sinister, the sponsorship of terrorists by states desiring surrogate vengeance or destabilization tools.

Nations facing threats by terrorism—most of the non-communist countries of the world but particularly the United States, Western Europe, and pro-Western Third World countries—have in the last two decades identified a need to develop capabilities targeted against those threats. The emphasis that has necessarily been placed in recent years on the upgrade of military and police forces worldwide oriented toward combating terrorism has been in one sense reassuring but in another somewhat saddening.

Military special operations forces have or can have a number of significant roles in the various functional areas of a national antiterror strategy (of which reaction, the most commonly imagined, is but one).

The National Strategy

The first element a national strategy ought to include is *problem identification*. What constitutes the major potential threats? What do we know about them? What is their casus belli? How do we arm against them?

Second is *co-option.* This includes domestic and foreign policy options that can remove a group's reason for striking a particular government or reduce or eliminate that group's appeal and potential public support. The Anglo-Irish Accords and Salvadoran land reform programs are examples of attempts at co-option (not necessarily successful).

Third is a collection of interrelated elements termed *proactive elements.* They include prevention through preparation, prevention through deterrence, and prevention through detection and preemption. They encompass organizing and planning to meet a threat at home or overseas, conducting the necessary prevention and response training for crisis management, detection, intelligence, and tactical personnel, and then ensuring a potential aggressor knows that the capability exists to bring his efforts to grief. There is not unanimity on this last point, but the statement of Doctor Strangelove, who said "you don't have a deterrent if nobody knows you've got it," is worth remembering. By the same token, detection can, but may not necessarily, lead to preemption; preemptive action by a principal or a surrogate depends on a variety of factors, including perception of cost-benefit ratio in terms of both innocent lives and political capital and confidence in the intelligence effort. But while it is true that detection may or may not result in preemption, it can at least result in interception or defense.

Finally, there are the *reaction elements:* relief and rescue operations for in-progress incidents and retaliation operations for incidents that have already occurred. The objectives of the latter elements, other than the obvious life-saving and punitive ones, are demonstrative, both to a national public and to other potential adversaries, that they may be assured of the government's ability and will to protect, rescue, and punish. Both of these elements also depend on reliable, real-time intelligence. Intelligence per se has not been mentioned as a functional strategy element, chiefly because it—along with national mood and governmental political will—forms the essential environment in which those functional elements exist. Without intelligence, the abstract capability to operate any element cannot become translated into meaningful policy options, as we saw in the case of the TWA hijacking in 1985 and have seen since then with regard to other Americans taken hostage in Lebanon.

These elements, then, and the capability to carry them out, form the basis of a national strategy. The additional ingredients of will to act and the informed decision making that implements that will and gives it direction, are what operationalize that strategy.

Special Operations Roles

Reaction and retaliation are the roles that spring most readily to the popular imagination when one considers SOF roles in the fight against terrorism.

Forces organized for these roles must be trained in a wide variety of weapons, equipment, operational scenarios, threat characteristics, and close combat techniques. A truly adequate training and preparation period is lengthy (and costly) and must be followed up with extensive simulation and field exercises if those skills are not to atrophy. That exercising and refresher and revalidation training may not require daily dedicated application to the exclusion of other missions, but many police and some military forces with counterterror or SWAT missions have them only as an adjunct to normal tasks and consequently can spend only limited time and resources on training. The inexperience of such forces and the dangers associated with their use in a delicate rescue situation inevitably manifest themselves in the few exercises they do run or in lives lost during actual incidents. Identifying personnel and buying some equipment does not translate into functional capability. Doctrine, and intense training oriented to that doctrine, does.

Another critical component in using SOF for such highly charged and politically sensitive operations as rescue and retaliation is the requirement for a working command and control structure. This includes not only a streamlined communications link from national command authorities but a combination of maturity, technical and tactical proficiency, and understanding of the incident environment—meaning the flow of intranational and international events affecting or potentially affecting an operation—on the part of those on "the sharp end." This may sound obvious, but consider some public-record revelations concerning the flexing of America's own counterterror muscle overseas.

At Desert One, only the army force involved was a self-contained, dedicated counterterror element. The air force and navy-marine aviation elements were not and for security reasons had not trained with the other actual participating elements. This ad hoc arrangement, failures to carry out basic premission inspections for such forbidden items as marked maps, photos, and agents' names, and the plan itself—a complex one chosen over several other feasible options—were all indicative of an inadequate command and control establishment. That was admitted by the government and addressed by the Pentagon in setting up the Joint Special Operations Command. No institutionalized, professional counterpart yet exists at the national command level, however, though many have been pushing for that for some time, and indeed until the Iran-contra scandal some semi-institutionalized crisis operations capability had been developed at the NSC. As it is, the signals from different sources—Department of Defense, Department of State, the NSC—can sometimes conflict. The case in point here is the TWA hijacking. A perceived window of opportunity appeared at one point in the plane's odyssey, U.S. forces were positioned to go, a go-ahead was apparently given, then rescinded, and while the confusion was resolved, the window closed, never to open again. And with regard to ground ledership, the precipitate decision to go ahead with the poorly intelligenced, hastily planned raid on Richmond Hill prison in Grenada in 1983, and the near debacle between U.S. forces and the Italian carabinieri at Sigonella air base in 1986—fairly accurately retold on an NBC special about the *Achille Lauro* incident—are cause for some concern. Yet whatever shortcomings may yet exist in U.S. counterterror forces' top leadership and national command and control links, the troops and the greatest majority of their leaders are highly skilled, highly dedicated, and utterly professional.

Less visible, less viscerally satisfying, but infinitely more important in the long run are the *co-optive* and *proactive* strategy elements. Co-option attempts to remove terrorism's raison d'être. Because most hard-core terrorists have generally adopted the means of their original cause as the foreseeable end, they cannot usually be successfully targeted, but their potential support mechanisms can. Where PSYOPS capabilities exist, they can usefully assist in spreading word and rationale for government actions, denouncing terrorists, protraying the inhumanity of "war on old people, women, and children," and so on. Appropriate behavior and civil and medical assistance operations by SOF elements in MTTs can materially assist in such co-option efforts as the grass-roots level. Conversely, personnel on MTTs who treat their charges and the civil populace with condescension and cynicism can only help enrich the terrorists' potential support base.

The proactive elements are those that, if implemented effectively, will seldom be evident to the population at large unless the government chooses, for political or civil morale-building purposes, to make them so.

The first, preparation, is the most broadly based in scope and the most difficult to measure in terms of effectiveness of implementation. In the United States, SOF are, have been, or could be involved in these several activities that go into preparation:

Civil threat awareness: SOF elements are involved here only to the extent that their masters choose to showcase response and prevention capabilities as part of overall discussions on the nature of the threat and the country's response.

Planning for prevention and response to attacks on critical facilities: SOF elements can, and have in the United States, participated in domestic multiagency contingency planning and exercises relative to critical energy and weapons production facilities, vulnerabilities, and recovery planning for U.S. embassies and major headquarters overseas and as standby intervention forces at both the 1980 Lake Placid Winter Olympics and the 1984 Summer Olympics in Los Angeles.

Antiterrorism training and assistance: Beyond the training that law enforcement and military forces do internally and with each other, there are international programs in which the military forces can contribute. These include protection training for diplomats, their drivers, and guards overseas; training for protection details and antiterror forces of other countries; and joint exercise training with major counterterrorists forces of one's allies. It has also included some pinpoint instances of assistance to other countries' forces during the resolution of incidents or participating as observers. U.S. SOF have performed in all these roles, as have the British, French, Israeli, and West Germans.

Such disparate elements of policy contribute to an overall strategy only if they proceed from centralized policy planning and occur under centralized direction, even if execution is decentralized. In the United States, a diversity of ad hoc committees, offices within the several agencies, and military commands share often overlapping responsibilities. The appropriate bridge point for fusing foreign and domestic-oriented forces, actions, policies, and intelligence actions is at the NSC, but thus far, policy and intelligence taskings have been decided by interagency representatives' votes and interagency planning and exercising done primarily through individuals' initiative in one or more agencies. The post–Iran-contra NSC bodes to be less, not more, inclined to institutionalizing and consolidating policy direction and operations in this area.

Another proactive element is deterrence. This is manifested in two ways: the example to be drawn by terrorists and those who support them, from the implementation of other elements such as recovery, preemptive, and retaliation operations, in all of which SOF have or can have a role, and in building awareness of capabilities specifically for the benefit of would-be attackers. There is divergence of opinion here, and not without reason. If all response and prevention capabilities and actions are kept secret, no deterrent factor can exist. But wholesale revelation of techniques, personnel, tactics, and movements can give adversaries who are determined to act information that can potentially be used against the counterterrorist forces. The Germans' GSG-9, the Austrians' Cobra Unit, and the U.S. FBI Hostage Rescue Team have all gone public in varying degrees, while the SAS continues to refuse public comment on its British counterterrorist plans and activities, and the U.S. Defense Department refuses even to acknowledge counterterrorist unit designations, though they have been matters of public record for years. This disparity in policy in the United States between Justice and Defense resulted in some loss of what had originally been a close working relationship between the Hostage Rescue Team and Defense's counterterrorist forces. Its existence can be traced only to noncentralized policy development and cannot be said to reflect any true approach to deterrence, unless it is to let the terrorists draw their own conclusions.

The trickiest proactive element is preemption. As with the other dynamic elements, this one, to be effective, is absolutely dependent on accurate intelligence, as well as such other mundane commodities as timing, positioning, and common sense. Preemption can take one of several forms. One is to conduct raids in response to discovery. This can range from relatively low-level police raids on safe houses or caches, in countries where police forces are cooperative, to full-scale over-border assaults on terrorist havens in unsympathetic areas. SOF can serve as participants (if they are police themselves) in the former and will figure very largely in the latter. Examples of the former include some excellent work done by Turkish, French, Italian, British, and Spanish paramilitary police forces in the wake of the 1986 U.S bombing attack on Libya, which effectively crippled a planned Europe-wide counteroffensive by Qaddafi's own operatives and other sympathizers. Examples of the latter are operations, undertaken overtly or covertly, by Israeli and South African commandos.

A second form is operations to minimize capability. These include raids on training camps, transportation centers, and means of sustainment in nations that harbor terrorists. SOF can be utilized in these roles, though they must generally operate covertly. The United States has operated these elements, though without use of SOF per se, in Libya and in the cigarette-boat contra raids on Nicaraguan refinery and port facilities.

Preemption can also include elimination, or "fingering" (intelligence collection) for later elimination, seizure, or turning of terrorists or terrorist supporters. The United States has thus far officially resisted calls to assassinate terrorist leaders, for a variety of reasons both moral and pragmatic, though there continues to be a call for it from several quarters. The United States does, however, train its SOF in a variety of legitimate target designation roles, and they could be utilized in that capacity if required. The SAS, the Israelis, and the South Africans have considerable experience in this area.

A possibly more palatable use is suggested in Secretary of State George Shultz's and President Reagan's assertions that the United States would act to have extradited, or, if cooperation were not forthcoming, would recover unilaterally known terrorists who have acted against Americans, to the United States for trial. This intention was put into practice in the Mediterranean "yacht sting" of the summer of 1987 (though not by SOF). Certainly SOF could be utilized in such actions—always presuming that the necessary intelligence is there and coordination is accomplished with affected allies. It was unsurprising, for example, that the U.S. counterterrorist force wanted to try to take Mohammed Abbas and the *Achille Lauro* terrorists prisoners when the plane they were on was forced to land at Sigonella, Sicily. But the base is under Italian legal jurisdiction except in cases covered under status of forces agreements. This instance was not such a case, and absent any high-level coordination, the Italian Carabinieri rightfully asserted themselves. The U.S. commander's understandable but unwise anger and heavy-handed threats nearly resulted in an intra-allied bloodbath.

Conclusions

So many roles, so many missions. Can any country's forces, particularly the smaller ones facing a budget squeeze, meet them all? One answer is that they do not all need to, principally because some countries do not have an overseas presence—either official or as embodied in tourists or businesses—significant enough to warrant measures or forces greater than those required to cope with subnational threats or transnational threats localized in or near national territory. But even for those countries with greater resources, the capabilities maintenance challenge looms large.

It is obvious that SOF have received renewed policy emphasis as a result of the terrorist threat. Translating policy emphasis into credible capabilities, however, depends first on having a clear strategy and then the will and resources to implement it.

8
The Special Operations Communications Challenge

Earlier we examined some of the communications issues facing U.S. SOF. To some degree, similar challenges face other countries' SOF. Here we will expand somewhat on the communications issues raised earlier, on the requirements that can be identified, and how current systems, new technology, and SOF procurement authorizations in the United States and United Kingdom, as examples, are meeting needs in this critical area.

Requirements

A principal consideration for any SOF field radio, as for any other item of equipment, is reliability. Regardless of capabilities, size, weight, and other characteristics, the equipment has to perform consistently in abusive operating conditions, transmitting and receiving over long distances. Weight and size of field use communications equipment are also critical factors. For long, independent reconnaissance or unconventional warfare operations deep into hostile territory, everything required for survival, operations, and communications must be carried (generally on foot) by the deployed special operations unit. Under these conditions, every pound carried must be a necessary one. Nor are the communications equipment packages themselves the only communications size and weight considerations: batteries, in quantities driven by anticipated use drainage, or hand-cranked generators, must also be packed to power each communications set. Thus, the search for smaller, lighter communications equipment has had a companion effort: a desire to minimize battery drainage through improved set design and improved batteries. Deployment and employment considerations go a good deal beyond reliability, size, and weight, of course.

In a reconnaissance or guerrilla base camp mode, the special operations unit may have to leave the radio unattended for periods of time. Rather than miss scheduled communications shots from the unit's rear control base, it is desirable to have a data-handling capability on the radio that can receive, store, and later provide a readout of incoming messages.

Another major concern is the radio's electromagnetic signature. All SOF operating in denied areas take great pains not to advertise their locations, yet they must carry on radio communications, one of the principal potential sources of compromise. Thus, there has always been a requirement for executing transmissions in coded form at the highest word-per-minute rate possible in the least time and for continuous frequency changes. One technique long used in Western SOF is receiving and transmitting on different frequencies at the outset. Historically, meeting these requirements (where they have been met) has beend done at the expense of size and weight considerations.

Another consideration involves antennas. Whereas conventional man-pack radios can be operated with a rod antenna connected directly to the set, the special operations unit usually needs to locate the radio in a secure, covert location and erect a larger, more efficient, static antenna. Unfortunately, secure locations for radios are not always the best locations for the antennas. The antenna terminating and turning unit should therefore be detachable from the main body of the radio and should itself be fully automatic and capable of silent or very fast tuning to reduce the probability of intercept. The rod antenna and backpacking capability must still be retained, however, and the radio equipped with voice communications capability because it may need to be utilized for interunit operations or, in some instances (generally, non-denied-area operations), secure voice communications with other headquarters.

Responses

Meeting the full range of their mission requirements with communications support has been elusive for Western SOF for a variety of reasons. Initially, neither the technology nor the perceived need supported any qualitative improvements over the World War II equipment utilized by the U.S. Office of Strategic Services and the British Special Operations Executive in their behind-the-lines espionage, sabotage, and resistance-support operations. Over the postwar years, spectrum occupancy and adversary electronic warfare capabilities increased but occasioned only minimal SOF radio improvements, such as the coder-burst device for the multiunit, hand-crank-generator-powered AN/GRC-109 in the United States. Some purpose-designed equipment was also developed in the United Kingdom and Europe, but those were designed primarily for use in Malaya and other end-of-empire conflicts then current, not for protracted, deep European operations.

Also retarding special operations communications advances has been the nature of special operations itself. Special operations equipment needs, compared to overall force procurement priorities in all Western armies, have long been seen as esoteric footnotes in the budget process. The density of requirement has always been so low that building any kind of constituency for SOF equipment modernization has been difficult, both with Western defense planning circles and within the production community. Until quite recently, no sense of urgency was accorded SOF equipment requirements in general or communications needs in particular. Requirements were validated but either unfunded, underfunded, or deferred.

The United States did proceed from the venerable 109 to a radio that could send and receive both code word and voice out of a single unit. The AN/PRC-74 was fielded in the early 1970s. It was not perceived as reliable as the 109 in the field, however, and it could not net with tactical voice systems. Most damning, from many users' standpoint, was the high battery drainage factor. The AN/PRC-74 has been replaced by the AN/PRC-70, intended to be "the" American SOF radio of the late twentieth century. However, the leisurely pace with which it was developed (seventeen years from design to field issue) reflected a continuous process of trying to stuff as many capabilities into an existing package as possible. The result is a hybrid, which does indeed represent some improved capabilities. With its digital message device group, it can communicate with base radio station GRC-122s or the newer TSC-99s. With a Vinson rig, it can communicate to a similarly equipped set hooked to secure FM, thus achieving the conventional tactical headquarters liaison capability long sought. It is reasonably reliable, but it weighs a tactically unacceptable 45 pounds and has high battery drainage.

In spite of these developments (such as they were) in the United States, it was not until the late 1970s that conscious decisions began to be made in the United Kingdom regarding development of an HF radio specifically designed for the special operations role. At that time, British SOF had the same budget priorities—almost none—as their U.S. counterparts, so while the requirement was validated, it was virtually unfunded. In the Falklands, as a result, the British SOF had no adequate issue communications equipment. What they did have was a few sets of the then new CALLPAC radio, developed by MEL, Inc., on loan. These were lifesavers, both in the immediate operational sense and in opening the door for funding further SOF communications development. The CALLPAC, while invaluable, did not have several capabilities long present on U.S. SOF radios, principal among them high-speed, short-duration, coded-burst transmissions.

Along with R&D funding, another constraint has been authorized stockage of existing equipment and manning of base signal support elements. In the United States, expansion of the army Special Forces and navy SEAL force structures has not been entirely matched by communsurate expansion of signal support equipment or operator manning authorizations. Thus, already thin base signal support assets are being stretched even thinner as the new units are forced to make do on equipment loaned to them by other units.

New Approaches

Although SOF will never assume budgetary primacy, despite rhetoric to the contrary, it is true that for most Western SOF forces, the stimuli of terrorism, actual force projections, increased long-range reconnaissance requirements, and, in the United States, declared Defense Department policy have all opened up the procurement checkbooks to SOF wider than ever. The time is ripe for exploring new communications options, and the United States and the United Kingdom are hard at it.

Both countries have developed satellite communications sets for

SOF use. These are most usable in a semistatic environment, such as a forward operational base, a training center in a counterinsurgency environment, or a semisecure patrol launch base. They have been used by teams in fluid, behind-the-lines operations, but that is not their optimum employment. The drawbacks of these sets are that, in any general conflict, loss of some or all the satellite platforms themselves must be presumed, and even in less-than-general circumstances, it is highly probable that preemption of available channels for more conventional intratheater or intertheater communications will greatly limit their utility to SOF. Recognizing that, both countries have development efforts underway that seek to meet the requirements identified for SOF.

One that has borne fruit has been development of smaller lithium (as opposed to standard magnesium) batteries with longer shelf life and a rechargeable nickel cadmium system. For field use, the lithium batteries have proved themselves, while the rechargeable systems have proved themselves primarily as annoyances; however, in protracted operations, such annoyances may have to become bearable. Other advances include those in the design of the United Kingdom's new PRC-319. The 319, in addition to having attractive size and weight factors (the receive/transmit unit weighs only 4.8 pounds, and the entire set with battery pack is only 12 pounds and measures 12 x 8 x 4½ inches), is equipped with detachable message units and an automated antenna tuner. The message unit has its own power source; thus, unit members at separate locations can punch out and store their own messages on the message unit, then return to the base location, reattach the message unit to the receive/transmit unit, and send the message. Operation of the message unit requires no operator knowledge of Morse code, and the burst capability is built into the radio. Also built in is a single-side band VHF capability. Its high power output (50 watts) can minimize or eliminate altogether the skip distance (space between the outer limit of groundwaves and inner limit of skywaves), which has traditionally bedeviled SOF communicators, and its capability of transmitting on one channel and receiving on another in 2 frequency Simplex on randomly selected channels increases survivability in an electronic warfare environment. The 319 relies heavily on state-of-the-art microprocessing technology and was designed from the ground up with special operations applications in mind. Similar capabilities in some areas can also be found on a U.S. prototype, the LORAL/SOUTHCOM MRR-50A.

An area only beginning to be explored is transmission systems based on laser media, which would defeat standard enemy intercept and decryption capabilities. Making laser technology amenable to all-weather, day-night general conflict SOF communications requirements involves a host of as-yet-unaddressed challenges.

Another technology that many in the SOF R&D community have found attractive is frequency hopping, or automatic changing of frequencies with each transmission and reception. This is currently being used in UHF and VHF fixed and mobile stations. At least one company in the United States, Harris, is developing a frequency-hopping HF radio for use by special operations forces.

Frequency hopping considered strictly as a function has some obvious advantages, principally in avoiding jamming, detection, and interception. In its implementation for SOF missions, particularly those conducted with or in proximity to allies, there are some considerations that must be addressed before it can be said to bear true utility. Among them is the general variance of frequency propagation strengths. A frequency that is good at 6 A.M. may not be (indeed, probably will not be) good at 6 P.M. Unless the hop-set covers the full range of frequencies, these propagation variances will adversely affect communications. And if the hop-set does cover the full range, it will (intentionally or unintentionally) wind up hitting channels utilized by friendly, enemy, and/or commercial and governmental sources. In short, it requires incredibly close, continuous frequency management, to a degree that may be practicable on a short counterterror operation but may not be in conjunction with SOF missions in Europe. Second, if the hop-set does cover the entire band, then a wide-band antenna is required. Currently these are much too large and bulky for SOF unit level. Conversely, if small, narrow-band antennas are used, then the hop-set is restricted, increasing the probability of intercept.

In general, though frequency hopping is an innovative approach and should be pursued for specialized application, it presents much greater frequency management, timing, and functional utility problems than do other HF-VHF options, which may be more suitable for the longer, more traditional SOF wartime missions.

The common denominator in SOF communications is perceived

need. Even if they see SOF utility only as deeper eyes, ears, and force multipliers for the theater commanders, conventionally minded planners have to some degree joined SOF advocates in recognizing that SOF can contribute but can do so only if they can communicate and survive to communicate again. Thus, diverging approaches to SOF utility can make common cause in the area of communications, and consequently, some real activity in the research, development, and acquisition field is underway. The only dangers are those of the activity losing momentum or of doing what the West has done all too often in SOF procurements—letting too many chefs spoil the soup. If the interest and activity in this most critical area is seen through to logical conclusions and aggressive procurement, then everyone wins: the SOF themselves, the conventional commands they support and complement, and the governments they serve.

9
Reflections and Conclusions

Popular imagination in the West has always been taken with special operations, whether those actually carried out or as contrived by writers of fiction and movies. The fascination has, to be sure, been somewhat ambivalent. On one hand, there has been the undeniable appeal of small groups of men bringing off tremendous deeds of daring in the face of daunting odds; on the other, the fear of "technicians of death" running out of control, or of being the dark, covert enforcers of unpopular or even illegal government foreign policies.

The course of recent history—Grenada, the Falklands, the numerous reductions of hostage and barricade situations in Europe, and the failed U.S. rescue raid in Iran, as well as the counterinsurgency operations in which many countries, including the United States and the Soviet Union, are currently engaged—has provided ample fuel for that imagination. And as if facts are not enough, fancy weighs in continually with new entries in the unprecedented volume of novels, television shows, and motion pictures in which current or former Green Berets, SAS, *spetsnaz,* or mercenaries are either good guys, bad guys, emotional cripples, or cool, amoral, sophisticated antiheroes.

Indeed, given the increased official attention (whether actual or for public consumption) and unofficial interest (whether sophisticated or sophomoric), one might almost lose sight of the fact that special operations' utility as policy options and special operations forces' utility as implementers of those options are nothing new. The Office of Strategic Services and Merrill's Marauders from the United States, the British Special Operations Executive and other numerous commando organizations, the resistance and sabotage experience of most of the rest of Europe, the Germans' Brandenburger Regiments, and the activities of the Soviet Union's partisans and SMERSH assassination squads (who actually existed and were not a figment of Ian Fleming's imagination) were all developed in World War II. The common thread in their origination was the recognition that disruption of enemy command and control facilities, lines of communication, and the integrity of certain strong point defense systems could materially impair the enemy's ultimate war-fighting potential, tie down troops who might be used elsewhere, provide invaluable intelligence, facilitate escape and evasion of downed pilots and compromised agents, and, by giving heart and stimulating further resistance on the part of occupied areas, weaken enemy political control.

That rationale for special operations—and troops trained and organized to carry them out—is no less valid in the late twentieth century that it was during World War II. Certainly many political and military leaders appear to have overlooked that fact at varying times in various countries' postwar histories, and one may be forgiven a certain amount of cynicism in listening to official expostulation of support for special operations made by those who considered them expensive and of nuisance value only a few years back. But three general conditions of current national and international affairs have given special operations and forces trained to implement them not only an unprecedented peacetime relevance but a general war-fighting importance beyond that of 1940–1945.

The first is the need to counter the increasingly sophisticated and lethal threat of transnational, subnational, and state-sponsored terrorism. As has been apparent throughout this survey, considerable resources have been dedicated by the United States and Western Europe to organizing and training dedicated military and/or paramilitary police forces to combat threats that manifest themselves both within and outside a country's borders and that are beyond what are considered normal police capabilities. The principal role for SOF in countering terrorism one might immediately consider is reaction. But there are a good many others, and these were addressed in some detail in chapter 7.

The second condition is the persistent fact that we live in a world of neither total peace nor total war but a nonetheless vigorous East-West competition that occupies varying positions within the shades of gray on the international conflict spectrum, a competition being fought out in the hills, jungles, and cities of the Third World. Outright conventional force intervention, either to crush an insurgency or to support one, is sometimes the only answer, and when that is the case, special operations perform critical reconnaissance, target designation, and selective strike functions. But more often such intervention is neither feasible nor politically palatable, and the preferred option is use of small teams of trainers and advisers in security assistance operations. Particularly for the United States, Great Britain, and France in the West, those small teams, manned largely by SOF personnel, are the much-preferred influence-projection option: better to assist a client state in defending itself than to be seen taking over the job from it. SOF are the first-team players in many aspects of this, the real war.

Finally, even in a general conflict, both for NATO and the Warsaw Pact, special operations have assumed main-event (versus sideshow) status. For the Pact, especially if surprise is maximized at the cost of a more force-rich but obvious buildup, Soviet and other special purpose forces operations to cripple the speed and sustainability of a NATO response through strikes on C^3I centers, leaders, and avenues of approach become not just important but critical. Moreover, while both *spetsnaz* operatives and those of the Soviet, German, and other regular intelligence services have undoubtedly already been over much of the ground around the Pershing, Lance, and ground-launch cruise missile bases, accurate, rapidly conducted target acquisition updates and designation operations will be essential to neutralize those threats. Compounding the challenge for the Soviets is the growing recognition in NATO of what real damage special purpose forces in its rear areas can achieve and the concrete steps being taken to harden *spetsnaz* targets. For the West, time for mobilization and intelligence are the two critical factors. Thus, SOF disruption and interdiction of enemy reinforcing echelons well in front of the forward line of troops is an essential component in overall efforts to halt the momentum of an enemy advance; and the eyes and ears of deep reconnaissance teams will not only complement but may almost entirely have to take the place of theater and strategic technical intelligence collection means.

There are, of course, varying levels of appreciation, and, particularly in the West, notable areas where reality sometimes does not match rhetoric. In the United States, SOF revitalization became something of a political football, with an impatient legislature forcing perceived organizational and command-control enhancements on an unenthusiastic Department of Defense and JCS. Even a year later, the senior Defense position created to oversee SOF programs had not been filled. And the remark of a senior officer in one of the new light divisions, created ostensibly to operate in low intensity conflict environments, who when advised that aggressor Special Forces had destroyed most of his helicopters stated it was just a nuisance and that his "battle focus" was the enemy tank capability, may speak volumes about where many of the U.S. military's conventional hearts and minds still are. Overseas several NATO countries have consistently added to their special operations forces' missions bag without significant additional resource allocations. These anomalies, and many more like it, are at odds with some very real improvements made in the United States and some other NATO forces and in interallied operations and planning. They cannot be dismissed and leave justifiable, lingering questions about commitment and true understanding. Doubtless there is grumbling about special mission prima donnas among Soviet senior officers, too—or at least when they are sure there are no GRU ears around.

But problems of conventional mentality, true understanding, political turf battling, and reality versus rhetoric are facts of life for the soldiers of the shadows of every country, as they always have been. What is different is that their missions—and their ability to perform their missions—are more critical now than ever. Most—not all, but most—military and defense planners recognize this, enthusiastically or otherwise. What they have not in all cases been as quick to recognize is the requirement to think through what they want special operations to achieve and then provide the material, moral, and political resources to support the forces involved. Only when that occurs can they be truly confident of their special operations forces' ability to perform their full spectrum of missions in peace, war, and the shades of gray.

Index

About the Author

Ross S. Kelly is currently the chief of the Defense Intelligence Agency's Instability/Insurgency Branch, where he is responsible for analysis and options development relating to low-intensity conflict situations. Previously he was a national security operations consultant specializing in the theory and practice of unconventional warfare, counterinsurgency, and other areas involving special military and paramilitary operations. His military experience includes command, intelligence, and operations positions in special operations forces and airborne infantry units and staffs. He served in Vietnam as an adviser to a Vietnamese paratroop battalion during the 1972 North Vietnamese Easter Offensive, and was awarded the Distinguished Service Cross for actions in combat. He writes and speaks frequently on special operations and low-intensity conflict.

Issues in Low-Intensity Conflict Series

Neil C. Livingstone, consulting editor

Other Publications in the Series

The Safe Travel Book
Peter Savage

Beyond the Iran-Contra Crisis
edited by Neil C. Livingstone and Terrell E. Arnold

The Handbook for Effective Emergency and Crisis Management
Mayer Nudell and Norman Antokol